BACTERIA AND PLANTS

THE SOCIETY FOR APPLIED BACTERIOLOGY
SYMPOSIUM SERIES NO. 10

BACTERIA AND PLANTS

Edited by

MURIEL E. RHODES-ROBERTS

AND

F. A. SKINNER

1982

ACADEMIC PRESS

A Subsidiary of Harcourt Brace Jovanovich, Publishers

LONDON · NEW YORK
PARIS · SAN DIEGO · SAN FRANCISCO · SÃO PAULO · SYDNEY · TOKYO · TORONTO

ACADEMIC PRESS INC. (LONDON) LTD.
24–28 OVAL ROAD
LONDON NW1 7DX

U. S. Edition published by
ACADEMIC PRESS INC.
111 FIFTH AVENUE
NEW YORK, NEW YORK 10003

British Library Cataloguing in Publication Data
Bacteria and plants. — (The Society for Applied Bacteriology symposium series; no. 10)
1. Plant diseases — Congresses
I. Rhodes-Roberts, M. E.   II. Skinner, F. A.
III. Series
581.2'3   SB727

ISBN 0-12-587080-9

LCCCN 81-68960

Phototypeset by Dobbie Typesetting Service, Plymouth, Devon, England
Printed in Great Britain by T.J. Press (Padstow) Ltd.,
Padstow, Cornwall, England

Contributors

D. B. ARCHER, *John Innes Institute, Colney Lane, Norwich NR4 7UH, UK*

J. E. BERINGER, *Soil Microbiology Department, Rothamsted Experimental Station, Harpenden, Hertfordshire AL5 2JQ, UK*

EVE BILLING, *East Malling Research Station, Maidstone, Kent ME19 6BJ, UK*

N. BREWIN, *John Innes Institute, Colney Lane, Norwich NR4 7UH, UK*

MARGARET E. BROWN, *Soil Microbiology Department, Rothamsted Experimental Station, Harpenden, Hertfordshire AL5 2JQ, UK*

J. G. CARR, *University of Bristol, Long Ashton Research Station, Long Ashton, Bristol BS18 9AF, UK*

M. J. DANIELS, *John Innes Institute, Colney Lane, Norwich NR4 7UH, UK*

CONSTANCE M. E. GARRETT, *East Malling Research Station, Maidstone, Kent ME19 6BJ, UK*

R. J. GILBERT, *Food Hygiene Laboratory, Central Public Health Laboratory, Colindale Avenue, London NW9 5HT, UK*

ILKKA HELANDER, *Department of General Microbiology, University of Helsinki, Mannerheimintie 172, SF-00280 Helsinki 28, Finland*

A. G. HEPBURN, *John Innes Institute, Colney Lane, Norwich NR4 7UH, UK*

A. W. B. JOHNSTON, *John Innes Institute, Colney Lane, Norwich NR4 7UH, UK*

BARBARA LUND, *Agricultural Research Council, Food Research Institute, Colney Lane, Norwich NR4 7UH, UK*

J. M. LYNCH, *Agricultural Research Council Letcombe Laboratory, Wantage, Oxfordshire OX12 9JT, UK*

M. J. M. MICHELS, *Unilever Research Laboratorium Vlaardingen, Olivier van Noortlaan 120, 3133 AT Vlaardingen, The Netherlands*

A. M. PATON, *Division of Agricultural Bacteriology, School of Agriculture, University of Aberdeen, Aberdeen AB9 1UD, UK*

T. F. PREECE, *Department of Plant Sciences, Agricultural Sciences Building, The University of Leeds, Leeds LS2 9JT, UK*

DIANE ROBERTS, *Food Hygiene Laboratory, Central Public Health Laboratory, Colindale Avenue, London NW9 5HT, UK*

RAGNAR RYLANDER, *Department of Environmental Hygiene, University of Gothenburg, S-400 33 Gothenburg 33, Sweden*

MIRJA S. SALKINOJA-SALONEN, *Department of General Microbiology, University of Helsinki, Mannerheimintie 172, SF-00280 Helsinki 28, Finland*

W. P. C. STEMMER, *John Innes Institute, Colney Lane, Norwich NR4 7UH, UK*

G. N. WATSON, *Food Hygiene Laboratory, Central Public Health Laboratory, Colindale Avenue, London NW9 5HT, UK*

Preface

A symposium on 'Bacteria and Plants' was the principal feature of the Summer Conference of the Society for Applied Bacteriology, held at the University College of Wales, Aberystwyth, 21–25 July, 1980. Contributions from many distinguished experts resulted in this volume which, in the opinion of the editors, constitutes a suitable text to extend formal lecture and laboratory courses. It should also provide a useful general background for research workers concerned with any stage between the growth of plants and the final consumption or processing of plant materials.

The first three chapters are concerned with bacteria associated with plant roots, either as inhabitants of the rhizosphere or as symbionts in nitrogen fixation within root tissue. Later chapters deal with plants overtly diseased in the field, with reference to both the initiation of infection and to its subsequent spread within the plant. Dissemination of bacteria between plants and the survival of infectious material between seasons are also considered. Special attention is given to the biology of crown gall disease because of the profound influence that elucidation of this tumorigenic process is likely to have on future work concerning the transfer of genetic material between prokaryotes and eukaryotes. Finally, the occurrence and activity of bacteria on plant materials is discussed in relation to the production of fermented products, spoilage of foodstuffs, public health hazards and the formation of toxic bacterial dusts.

The editors wish to record their appreciation of the keen interest taken in this Symposium by Professor J. G. Morris and Professor P. F. Wareing FRS, both of the Department of Botany and Microbiology in the University College of Wales at Aberystwyth.

MURIEL E. RHODES-ROBERTS,
Department of Botany & Microbiology
University College of Wales,
Aberystwyth SY23 3DA,
Dyfed

F. A. SKINNER,
Soil Microbiology Department,
Rothamsted Experimental Station,
Harpenden AL5 2JQ,
Hertfordshire

Contents

Interactions between Bacteria and Plants in the Root Environment

J. M. LYNCH

Agricultural Research Council Letcombe Laboratory, Wantage, Oxon, UK

Contents

1. Introduction

IT IS entirely appropriate that as plant physiologists further consider the importance of roots and their associated microflora with respect to plant growth (Russell 1977) we should start this symposium 'below stairs'. In this review I will attempt to explore whether some of the excitement engendered by the more recent studies of bacteria–root interactions is justified, and indeed whether such studies have potential relevance to agriculture. But before this can be done, it is first necessary to describe the anatomical, physiological and ecological aspects of the rhizosphere and point out some gaps in our knowledge.

2. Anatomy of the Rhizosphere and Microbial Community Structure

The rhizosphere was first defined by Hiltner (1904) as the zone of stimulated bacterial growth around legumes resulting from the release of nitrogen compounds by nodules. This description has since been modified and made more general; now it seems reasonable to include all microbial growth using root-derived compounds as sources of carbon, nitrogen and energy. The root epidermis–cortex zone, when colonized by pathogens or non-pathogens (Darbyshire & Greaves 1973; Old & Nicholson 1978), has been termed the 'endorhizosphere' by Balandreau & Knowles (1978). It therefore appears reasonable to describe the zone of colonization outside the root as the 'ectorhizosphere'. However it must be recognized that the zone where the ecto- and endorhizosphere merge is probably the region where some of the most intimate plant and microbial associations occur (Elliott *et al.* 1980) and this is termed the 'rhizoplane'.

The methods used to examine the rhizosphere have been reviewed (Lynch 1981). The ectorhizosphere and rhizoplane have been viewed by light

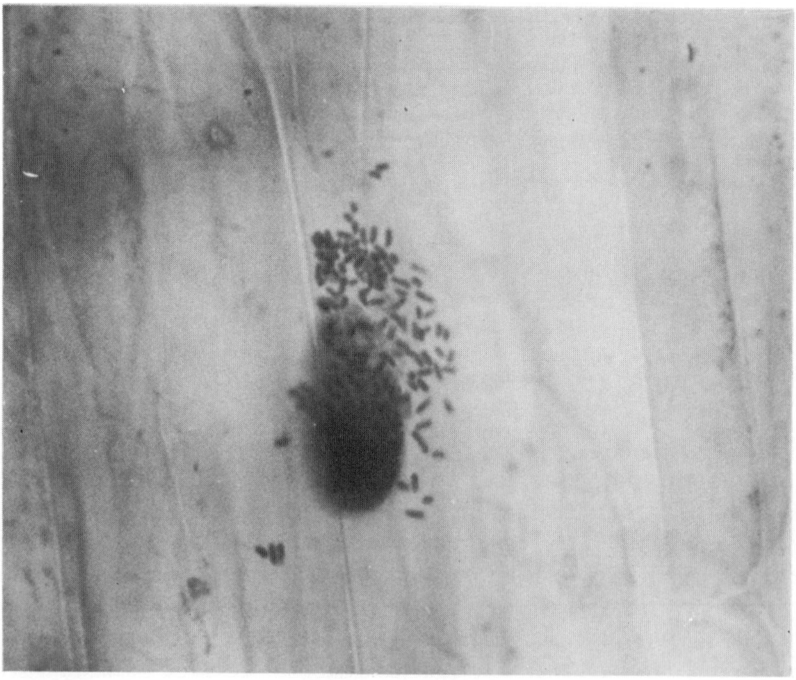

Fig. 1. Light micrograph of *Pseudomonas* sp. as a distinct colony on the surface of barley (*Hordeum vulgare*) root, × 2300. (Photograph by R. A. Bennett.)

microscopy usually after staining with phenolic aniline blue, but recently techniques involving fluorescence have been widely employed. In our experience (Bennett & Lynch 1981) the bacteria often occur as distinct colonies, with about 10^2 cells per colony, on the plant root (Fig. 1) and are often concentrated at the junctions between epidermal cells (Fig. 2). This non-random distribution of bacteria in the rhizoplane has been analysed mathematically by Newman & Bowen (1974) but it may be more appropriate to use more modern types of statistical analysis, as has been used for micro-organisms colonizing soil particles (Polonenko *et al.* 1978). We have found great difficulty in counting such closely aggregated cells and have not attempted numerical analysis. Assessment of the population size is also dependent on all the bacteria adhering to the root surface and we have observed great variations in the ability of different bacterial species to adhere.

The non-uniform colonization patterns have also been demonstrated with scanning electron microscopy (s.e.m.) (Figs 3 & 4), and s.e.m. has been used to examine the rhizoplanes of field-grown plants (Elliott *et al.* 1980). This

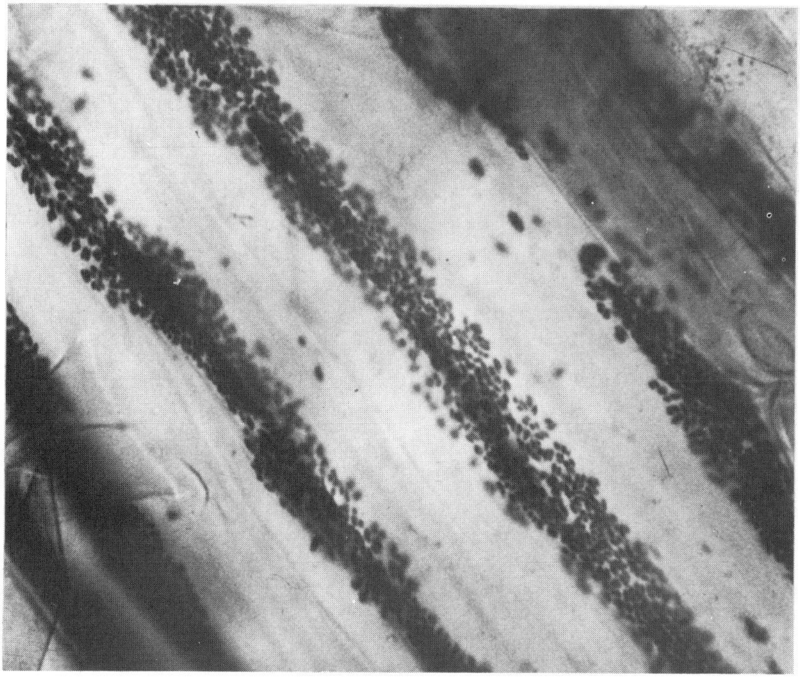

Fig. 2. Light micrograph of *Pseudomonas* sp. along the cell junctions of barley (*Hordeum vulgare*) root, × 2230. (Photograph by R. A. Bennett.)

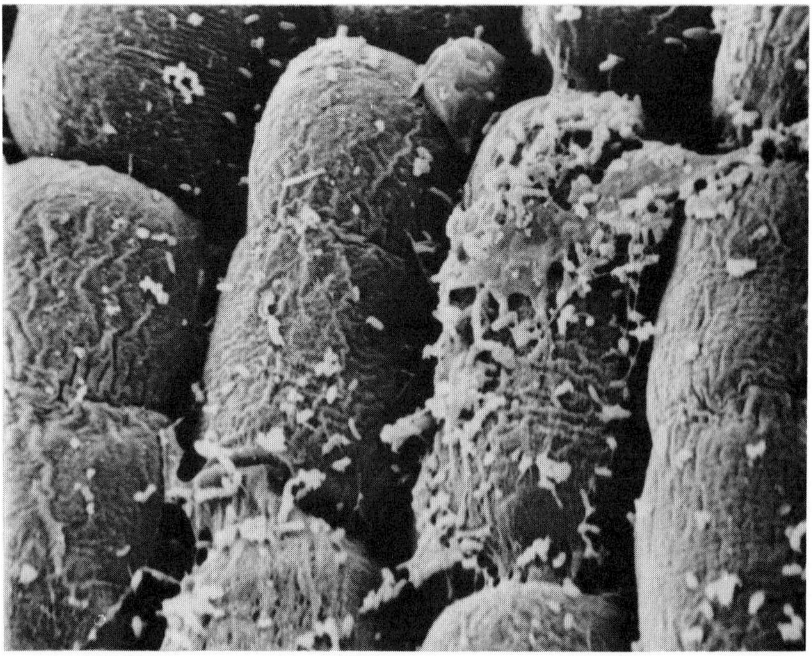

Fig. 3. Scanning electron micrograph of bacteria on the surface of maize (*Zea mays*) root showing sparse colonization, × 2030. (Photograph by M. C. Drew & R. Campbell.)

demonstrated that the rhizoplane of direct-drilled winter wheat plants had a greater microbial population than plants drilled into tilled land, but this did not affect the crop yield.

Transmission electron microscopy (t.e.m.) of transverse sections has been particularly useful in depicting the endorhizosphere, but it is very difficult to use t.e.m. quantitatively and therefore to assess the role of the endorhizosphere microflora in relation to the rhizoplane and ectorhizosphere organisms. Certainly there is a greater microbial biomass in the older parts of roots. The much closer proximity of the endorhizosphere bacteria to the conducting elements (xylem and phloem) of the plant mean that they are spatially better placed to interrupt the plant's nutrient supply and introduce phytotoxins or growth regulators to the translocation stream. Figures 5 & 6 show cortical cell wall decay by endorhizosphere organisms which include pseudomonads (M. P. Greaves, pers. comm.). Such cellular decay in crop plants leads to death of plants or at least to reduction in yields. We therefore need to investigate the ecology and physiology of the endorhizosphere which has so far received scant attention.

Microscopic examination has commonly demonstrated that bacteria

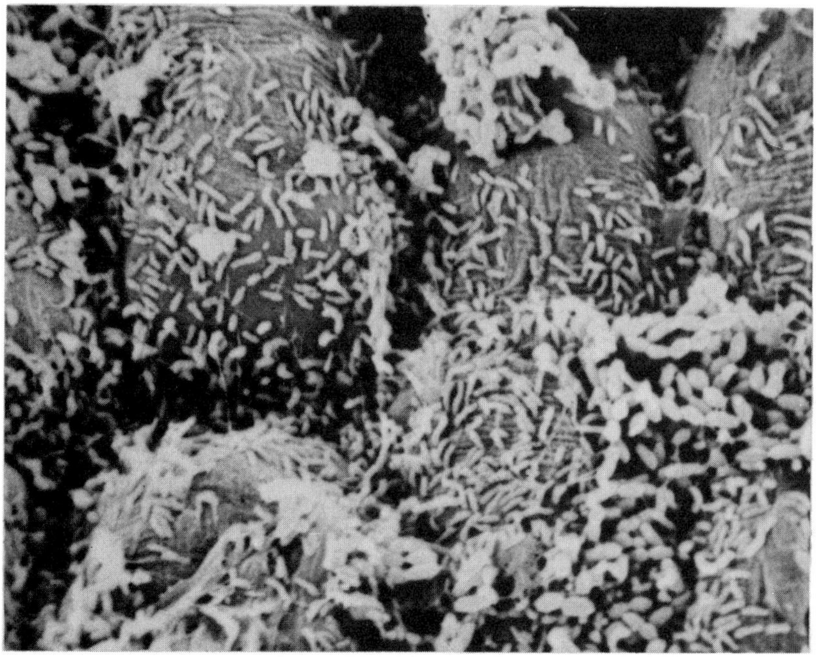

Fig. 4. Scanning electron micrograph of bacteria on the surface of maize (*Zea mays*) root showing dense colonization, from an area close to that in Fig. 3, × 2030. (Photograph by M. C. Drew & R. Campbell.)

predominate over fungi in the rhizosphere, but Newman *et al.* (1981) have shown that in a survey of 40 sites in England and Wales, the fungal to bacterial ratios, expressed as percentage cover of the root surface of *Plantago lanceolata*, varied from 0·28 to 14·0. Vancura & Kunc (1977), in an attempt to quantify the rhizosphere microflora by measuring respiration after treatment with selective antibiotics (streptomycin to inhibit bacteria and actidione to inhibit fungi), showed that bacterial respiration was the greater in the rhizosphere. This contrasts with the predominance of fungi over bacteria, on a weight basis, in the soil biomass as a whole. A special situation obtains when ectomycorrhiza develops. As a result of a specific, intimate association between the root and a fungus, the latter envelops the root as a sheath and plays a part in translocation of nutrients and water to the host.

As a root grows through the soil it will continually encounter sites occupied by different species of soil micro-organisms, yet there are claims that some bacteria, such as *Pseudomonas* spp., dominate the rhizospheres of particular soils (Rouatt & Katznelson 1961). What then governs the success of a rhizosphere colonist? The ability of micro-organisms to adhere to the rhizoplane may be one factor, but other than the interactions of *Rhizobium* with legume root hairs and the attachment of *Agrobacterium* to roots, both being mediated by the extracellular polysaccharides of the root and the bacteria (Dazzo 1981), there is little evidence in support of a similar mechanism for non-symbiotic micro-organisms. Presumably some organisms will exhibit chemotaxis in response to specific nutrients released by the roots; for example there is the attraction of the pathogenic fungus *Phytophthora infestans* to tree roots which release ethanol by fermentative metabolism in wet anaerobic soils (Allen & Newhook 1973), but I know of no similar observations with bacteria. Clearly these topics are important to our understanding of rhizosphere ecology and now that the necessary techniques are available to facilitate study, this should be considered as a research priority. The older data on the composition of bacterial species isolated from the rhizosphere may merely reflect the use of media and techniques now known to be unsuitable and indeed highly selective. It would be surprising if a single species ever dominated the rhizosphere, it

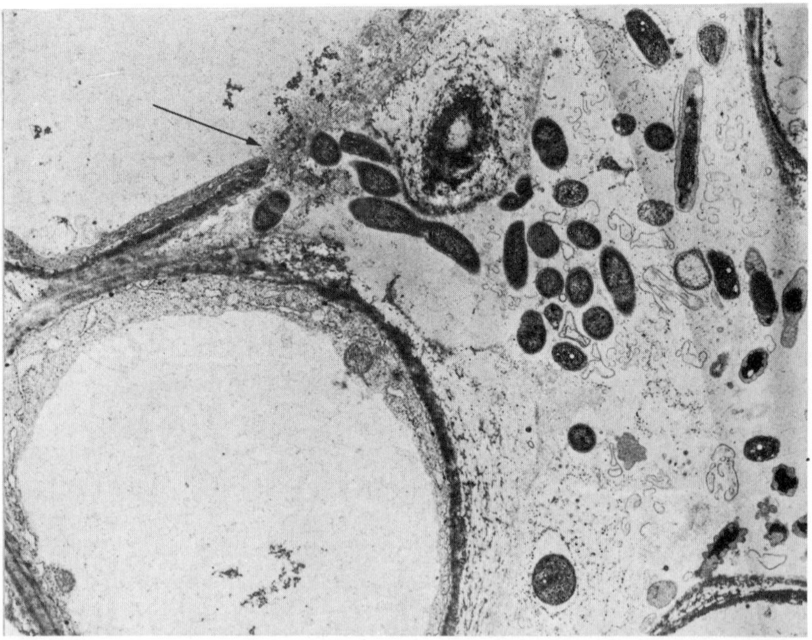

Fig. 5. Bacteria colonizing the intercellular space of the root cortex of wheat (*Triticum aestivum*). The normal 'in fill' of polypectate has been degraded and the penetration of a cortical cell wall (arrowed), three cell layers into the cortex, is evident, × 8385. (Photograph by M. P. Greaves & J. A. Sargent.)

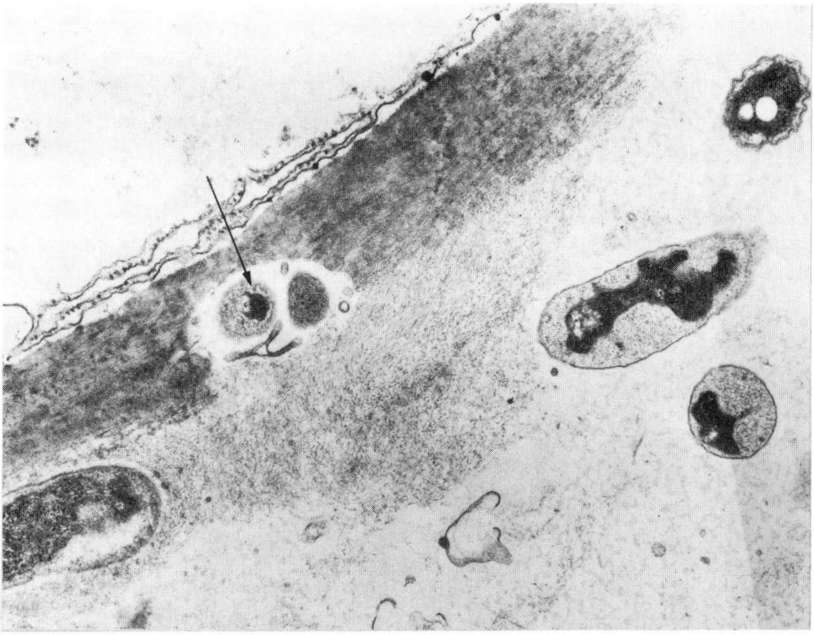

Fig. 6. Similar effect as that shown in Fig. 5 but bacteria penetrating the fourth layer of cortical cells (arrowed), × 30,780. (Photograph by M. P. Greaves & J. A. Sargent.)

appears more likely that a community will build up and become successful by co-operative metabolism. How long such a community can remain stable is again uncertain.

The density of colonization increases on the rhizoplane of the older parts of the root; this is probably a consequence of the greater release of substrates here and also the fact that the growing tip is not rapidly colonized.

3. Provision of Substrates for Microbial Growth in the Rhizosphere

The organic materials released by roots have commonly been referred to as 'exudates' but our terminology should now be more precise. 'Exudates' leak from living roots, whereas 'secretions' are actively pumped, and 'lysates' are passively released from roots during autolysis. 'Mucigel' arises from plant 'mucilage' and is produced from epidermal and root cap cells. Microbial growth around roots, using exudates, secretions and sloughed cells as substrates, leads to 'microbial mucilages' which also form part of the mucigel.

Early investigations on these root-derived substrates using aseptically grown plants showed that they accounted for only a few per cent of the plant's dry matter production. However more recently it has been possible to grow plants in non-sterile soil with $^{14}CO_2$ as the sole source of C available to the plant, then to fractionate the ^{14}C in the plant and the soil, and in the CO_2 respired by roots and soil micro-organisms, and hence produce a carbon balance sheet (Barber & Martin 1976; Sauerbeck & Johnen 1977). These investigations showed that about 20% of the dry matter produced by the plants was released by roots, but only about half of this was released into sterile soil. The release of C from roots in the presence of bacteria can also be estimated indirectly, by comparing that released in carbon compounds under aseptic conditions, with the carbon contained in bacterial cells (provided with no other carbon substrate) that are growing around the roots (Barber & Lynch 1977). This again demonstrated that bacteria appeared to stimulate the release from roots of substrates, the amount of which increased as the root aged.

The precise chemical nature of the carbon compounds released by roots is difficult to determine, but as would be expected, they comprise all the kinds of materials that are found in the cells (Rovira 1965). It would appear that carbohydrates are the major components and that the C/N ratio is about 30/1 (based on the amino acid content), which is much greater than that of microbial cells (Barber & Gunn 1974). Hence microbial growth in the rhizosphere will tend to immobilize soil or fertilizer N. However a full carbon and nitrogen budget of the rhizosphere has never been attempted and any detailed assessment of even the major constituents is very difficult because under non-sterile conditions, when the maximal amount of substrate is released, most of it will be rapidly incorporated into microbial cells.

4. Microbial Growth Kinetics and Mathematical Models of the Rhizosphere

The rate at which plants release substrates to the rhizosphere microflora is uncertain, but there are gradients of release, with maximal amounts being produced around the zone of cell elongation behind the root tip where lateral roots are formed (Schippers & van Vuurde 1978). Generally the size of the rhizosphere population per unit weight of root increases with the growth and age of the roots (Fig. 7). If, in this situation the substrate is released at a constant rate, the microbial population might be considered to be a 'fed batch culture' which has growth kinetics equivalent to a 'continuous culture' (Pirt 1975). Hence:

$$\begin{array}{ccccc} \text{Overall rate of energy consumption} & = & \text{Consumption for growth} & + & \text{Consumption for maintenance} \\ \dfrac{\mu x}{Y} & = & \dfrac{\mu x}{Y_g} & + & mx \end{array}$$

where μ is the specific growth rate (per h), x is the biomass (g), Y is the observed growth yield (g dry weight/g substrate), Y_g is the true growth yield when no energy is used in maintenance (g dry weight/g substrate) and m is the maintenance coefficient (g substrate/g dry weight/h). This is of course a gross over-simplification of the likely natural situation but it has provided the basis for a mathematical model of rhizosphere growth (Newman & Watson 1977). In the model the change in S, substrate concentration (μg/cm^3 soil), with time, t, in the rhizosphere/rhizoplane region was analysed, considering the root as an idealized cylinder, and expressed as:

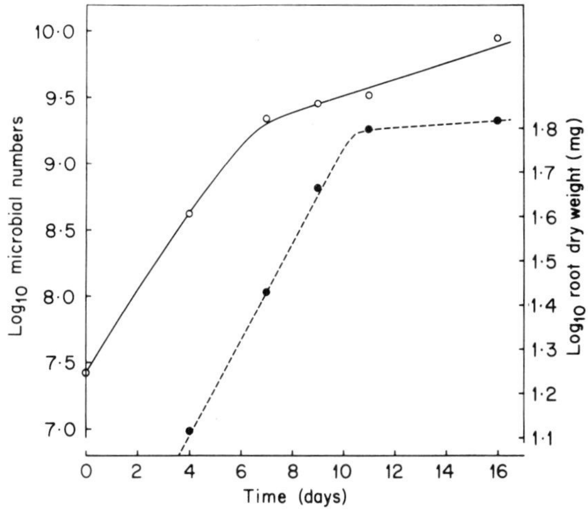

Fig. 7. Growth of barley roots and associated population of bacteria. o——o, micro-organisms; •- - - -•, roots. (From Barber & Lynch 1977.)

$$\begin{array}{ccccccccc} \dfrac{ds}{dt} & = & F_D & + & F_I & - & (F_G & + & F_M) \\ & & \text{(substrate to a radial distance, } r\text{, from root axis)} & & \text{(indigenous supply of substrate from soil)} & & \text{(substrate used in microbial growth)} & & \text{(substrate used in microbial maintenance)} \end{array}$$

The model was tested to assess the influence of some of the individual variables. Of the limited amount of data tested with the model, reasonably good fits were obtained. This finding should provide a stimulus to test the model further. The values used for the microbial maintenance coefficient need particular attention since we know very little about maintenance energy at slow growth rates. The amount of substrate available is considerably greater than that in the bulk soil but the rhizosphere population nevertheless grows only at a very slow rate (0·007–0·03/h) even in highly favourable conditions on plants growing in culture solutions in the laboratory (Barber & Lynch 1977). It is of interest that bacteria on the surfaces of aquatic plants in rivers also grow at a similar slow rate (Hossell & Baker 1979). These rates are much slower than those obtained by bacteria growing in culture media (up to 2/h); systems which have been used to derive values for maintenance coefficients.

5. The Chemostat as a Rhizosphere Model

That the kinetics of microbial growth in the rhizosphere might approximate to those of continuous culture has stimulated Coleman *et al.* (1978*a*,*b*) to use the chemostat as a model for the rhizosphere. The population used in their studies was a mixture of *Pseudomonas* sp. (a true rhizosphere inhabitant), *Arthrobacter* sp. (an autochthonous or indigenous soil bacterium), *Acanthamoeba* sp. (a protozoan predator of bacteria) and *Mesodiplogaster* sp. (a nematode predator of protozoa and bacteria); the carbon substrate provided was glucose. This provided some interesting and useful information on the interactions between the species studied; e.g. the *Pseudomonas* grew faster than the *Arthrobacter*, confirming what we suspected from observations under natural conditions. However, in order to build on these findings it seems essential to use stable communities which can be isolated from nature (Slater & Bull 1978). The same consideration applies to substrates; thus plant polysaccharides would probably be more suitable than free sugars.

6. The Role of Bacteria in Ion Uptake by Roots

Work in this laboratory and the CSIRO Division of Soils in Adelaide (for a review see Barber 1978) has demonstrated unequivocally that when plants are grown in culture solution under gnotobiotic conditions, bacteria influence the uptake of ions. Under some conditions ion uptake by plant roots can be stimulated by bacteria, possibly by providing chelating agents or plant growth regulators to promote active ion transport, but under other conditions they can be inhibitory, either by competing for the nutrients or producing phytotoxic compounds. These observations therefore underline the difficulty in interpreting plant physiological studies in laboratory culture solutions where the role of bacteria is disregarded. Phosphate uptake by plants, for example, has received great attention; phosphate occurs in the soil solution in small concentrations and is readily incorporated by bacteria into nucleic acids. Over short (30 min) periods phosphate uptake and translocation within the plant is promoted by bacteria, but over longer (24 h) periods uptake is reduced (Barber *et al.* 1976). Seedling age is also important, phosphate uptake being stimulated by the bacteria in younger (6 day), and inhibited in older (12 day), seedlings.

Despite the demonstrations of the role of bacteria in ion uptake by roots from plant culture solutions, there have been remarkably few investigations of their effects in soil. Such experiments can be performed in radiation-sterilized soil, and using this technique, Benians & Barber (1974) demonstrated that the presence of bacteria decreased the P content of plants. This effect however probably depends on the soil type and bacterial species present. For example, some bacteria solubilize P from soil minerals (Gerretsen 1948; Duff *et al.* 1963) and this could make phosphorus more readily available for plant growth in some soils. Odunfa & Oso (1978) have shown that more phosphate-dissolving bacteria occurred in the rhizosphere of cowpea than in that of maize. Thus some bacteria may provide plants with P, a function usually considered to be associated with mycorrhizal fungi. However the overall role of bacteria in P uptake by soil-grown plants remains unclear and warrants further study.

7. Competition between Bacteria and Roots for Oxygen

Although the ratio of microbial to root biomass is quite small (*ca.* 1:14), micro-organisms are well placed spatially to compete for available oxygen, which could become critical when there is a shortage, as in waterlogged soils. Although Griffin (1968) and Armstrong (1979) considered the effects of a limited supply of oxygen on roots and micro-organisms, no model has been published. Such a model should consider essentially the consequence of a cylinder of micro-organisms around roots (Drew & Lynch 1980). However this would be difficult to describe accurately because oxygen flux through the surface of a healthy root can be maximal in the apical zone where microbial colonization is minimal. Furthermore, although the requirement for oxygen (Q_{O_2}) by roots is usually much less than that required by multiplying micro-organisms, it is highly variable for different

TABLE 1
Oxygen uptake by bacteria and plant roots

	Q_{O_2} µl/mg dry wt/h	Reference
Azotobacter vinelandii	5000	Williams & Wilson (1954)
Escherichia coli	200	Harrison (1972)
Sterile plant roots	4	Lee (1980)

microbial species (Table 1). The *Azotobacteriacae*, which have high oxygen requirements, also produce extracellular polysaccharides which could further impede the diffusion of oxygen to roots. No experiments have been conducted with roots to test this hypothesis but we have demonstrated that *Azotobacter chroococcum* can inhibit germination and early seedling growth of barley (Harper & Lynch 1979). Exudates were released from the seed through a fibrous plug at the end containing the embryo. The plug then became the site for microbial colonization and it was also the pathway of oxygen entry into the seed. As the oxygen became depleted from the system, the seed released more exudates, thus promoting more bacterial colonization and a greater relative utilization of oxygen by the micro-organisms; eventually the seed sometimes died due, presumably, to anoxia. This effect only occurred when the *Azotobacter* strain used had been grown on a combined N source; those bacteria which were actively fixing N_2 never inhibited germination and sometimes accelerated it.

8. Provision by Bacteria of Growth Regulators and Phytotoxins to Roots

Bacteria have the ability to produce metabolites which in suitable plant bioassays show activity similar to auxins, gibberellins, cytokinins and ethylene (Brown 1974; Lynch 1976). With the exception of ethylene, few of these metabolites have been identified unequivocally. Furthermore, there are few studies describing the effects of adding growth regulators exogenously to roots growing under aseptic conditions. Brown (1972) demonstrated that shoot growth was promoted, but root growth inhibited, by adding a mixture of 10^{-9} mol/l gibberellic acid Type 3 (GA_3) and 10^{-8} mol/l indolyl 3-acetic acid (IAA); we also have observed inhibition of root growth by application of these growth regulators to seeds or roots, but we have not observed the stimulation of shoot growth (Lynch & White 1977; Harper & Lynch 1979). This aspect of bacterial metabolism would probably be more significant to the plant if the bacteria producing the metabolites inhabited the endorhizosphere. Certainly the production of IAA by rhizobia (Kefford *et al.* 1960) and mycorrhizae (Slankis 1973) is thought to affect plant metabolism.

Two major pools of substrates, crop residues and the organic carbon compounds from roots, are available for heterotrophic microbial activity in soil (Lynch 1979). Several studies concerning the role of crop residues as substrates for the formation of phytotoxins have been made. There is also a potential for root-derived carbon compounds to serve as substrates for micro-organisms and as the products are likely to be formed in close proximity to the root surface they are probably more significant to the plant. Such products are commonly formed only under anaerobic conditions in wet soils; this environment also favours their accumulation because degradation by aerobes is inhibited (Drew & Lynch 1980).

One of the few rhizosphere processes that has been studied in this context is sulphate reduction. The promotion of heterotrophic microbial activity by the release of carbonaceous materials from roots increases oxygen consumption in the rhizosphere, and as supplies of oxygen become limiting, the redox potential will decrease: if the potential reaches *ca.* −220 mV, sulphate-reducing bacteria can use sulphate as a terminal electron acceptor and hydrogen sulphide is formed. This activity has been observed in the rhizosphere and spermosphere of plants (Dommergues & Jacq 1972; Garcia *et al.* 1974). In the presence of excess Fe^{2+}, the formation of insoluble ferrous sulphide usually detoxifies the hydrogen sulphide (Gambrell & Patrick 1978) but in conditions where it persists, concentrations of *ca.* 0·1 μg/ml water inhibit respiration and poison cytochrome oxidase (Allam & Hollis 1972). The consequent damage that can occur to rice shoots is a serious economic problem, the syndrome being known in different parts of the world as browning, bronzing, brusome, *akiochi*, *hie-imochi*, suffocation and straighthead diseases (Hollis *et al.* 1975). Control is usually achieved by increasing the redox potential by applications of nitrate, but under natural conditions some degree of control may be achieved by rhizosphere inhabitants such as *Beggiatoa* in localized niches where hydrogen sulphide is oxidized extracellularly to sulphur (Joshi & Hollis 1977).

There are many other microbial metabolites such as aliphatic and aromatic acids which we know to accumulate in anaerobic soils and probably also to originate in the rhizosphere, but there has been little specific study of their formation in this region.

9. The Nutrient Film Technique

In the glasshouse industry the technique of growing crops, particularly tomatoes, in a shallow layer of nutrient solution circulating between two layers of polyethylene sheet has attracted great interest in recent years (Winsor *et al.* 1979). The advantages of this method are that it gives control of the root environment, watering is greatly simplified, uniformity in the supply of nutrient and crop protection chemicals is ensured, root temperature can be raised when necessary and the soil problems of salinity, poor structure and drainage are avoided; soil-borne pathogens are likewise eliminated, thus avoiding the need for regular soil sterilization and longer cultivation, and enabling a more rapid succession of crops. Crop yields have been very satisfactory, but it has always been recognized that microbial growth in the nutrient films could be a major problem. In practice there have been surprisingly few bacterial or fungal pathogen problems with tomatoes and it could be that a stable rhizosphere microflora which effectively resisted invasion by pathogens was created, analogous to that formed in the solution culture studies involving ion uptake described above. As the technique finds greater application it will be essential to study aquatic rhizospheres in more detail against the obvious possibility that a pathogen entering the system could rapidly devastate the crop. An increased understanding of rhizosphere physiology is highly relevant to this technique. To obtain maximum benefit, nutrient and oxygen consumption by bacteria should probably be minimized.

10. Are Non-symbiotic Bacteria–Root Interactions of Agronomic Significance?

By releasing 20% of its dry matter production to the soil the plant is apparently wasting energy. Although this process may serve a useful purpose, for instance by releasing toxins, no physiological role has been clearly identified; for the plant perhaps the process is non-essential. However, the process may modify the soil environment in the following possible ways.

A. The nitrogen cycle

This will be mentioned only briefly here as it is considered fully in the succeeding two contributions. It is however worth emphasizing that the carbon compounds released by roots may stimulate the growth of heterotrophs in the nitrogen cycle, including nitrogen-fixing and denitrifying bacteria. Associative (rhizocoensic) nitrogen fixation has been the subject of much speculation and discussion. We have calculated that if all the carbon released by roots were available only to known nitrogen-fixing bacteria, and if the nitrogenases of the bacteria functioned at their maximum rates, then only 15% of the N content of temperate cereals could be provided in this way (Barber & Lynch 1977). In practice of course, the amount would be much less and no protagonists of associative nitrogen fixation have yet fully defended their case. It may be that such an association would be significant for survival in nutrient-poor situations such as sand dunes. The argument that plants using the C_4 photosynthetic pathway would release more carbon than those using the conventional C_3 pathway, thereby providing more energy for fixation, has not yet been substantiated. However our search for a natural or engineered association should continue (Elliott *et al.* 1980).

It is clear that rhizosphere processes contribute to denitrification, whereby N is lost from the soil plant system as N_2 and N_2O (Smith & Tiedje 1979). The contribution of this in relation to the nitrogen economy of soils has not been clearly assessed but the loss of N_2O may be quite small (Burford *et al.* 1979).

B. Contribution of the rhizosphere to the soil biomass

It would be tedious and indeed unreliable to assess the ratio of the microbial biomasses in the rhizosphere and the soil by microscopic methods. Plate counts of bacteria are even more unreliable as they are so selective. However the introduction of the fumigation/respirometric method of soil biomass determination by Jenkinson & Powlson (1976) has made it possible to assess indirectly the rhizosphere contribution to the biomass. We have modified their procedure such that intact cores of soil are used (Lynch & Panting 1980*a*) and have shown recently (Lynch & Panting 1980*b*) that on adjacent plots, the microbial biomass is much greater in grassland than in arable plots (Table 2). Jenkinson & Powlson (1976) also found large

TABLE 2

Root weight and microbial biomass in the surface 5 cm of a clay soil (Denchworth series) under grass and arable cultivation

	Root weight (mg C/100 g dry soil)	Microbial biomass (mg C/100 g dry soil)
Grass	227	206
Oil-seed rape	8	72

Data from Lynch & Panting (1980*b*).

microbial biomasses in grassland soil. This is strongly suggestive but does not prove that the increased microbial biomass is a consequence of the presence of roots.

C. Effects of the rhizosphere on soil structure

Grassland soils generally have very stable soil structures (Low 1955; Robinson & Jacques 1958; Clarke *et al.* 1967; Reid & Goss 1980) and this may be associated with the larger microbial biomass. The chemicals responsible for the adhesion of soil particles and soil crumb stability with fluctuating water content are at present unknown (Burns 1981; Fletcher *et al.* 1981) but it seems that polysaccharides are at least partially involved. A comparison of the monomer units of polysaccharides of plants and micro-organisms shows that both contribute to soil polysaccharides, but the relative contribution of each is still uncertain (Cheshire 1979). Figures 8 & 9

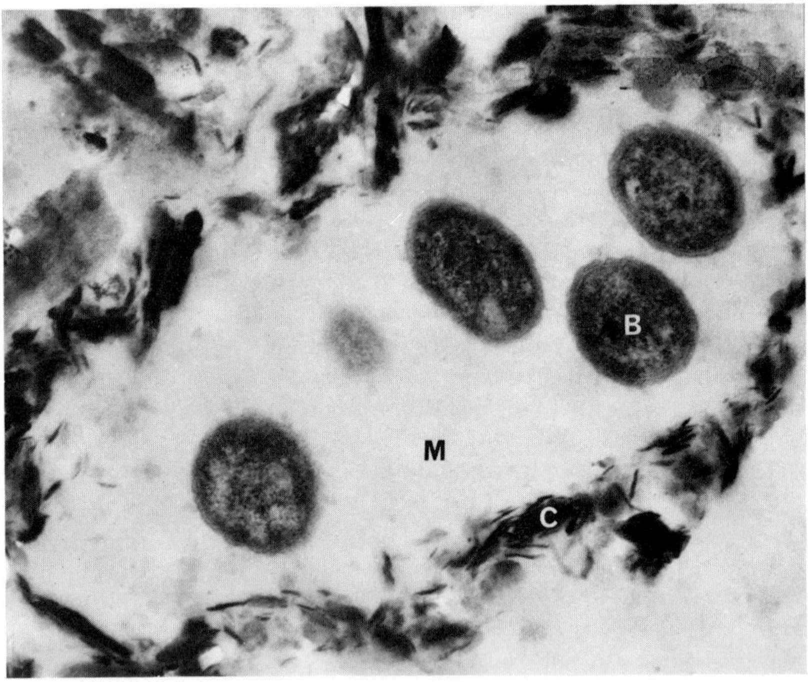

Fig. 8. A microcolony of *Bacillus mycoides* on the root surface of wheat *(Triticum aestivum)*. The bacteria are embedded in an electron-transparent material, probably mucilage, bounded by clay particles. M, bacterial mucilage; C, clay particles; B, bacterial cell, × 25,000. (Photograph by J. L. Faull & R. Campbell, reproduced by permission of the National Research Council of Canada from the *Canadian Journal of Botany* (1979) **57**, 1800–1808.)

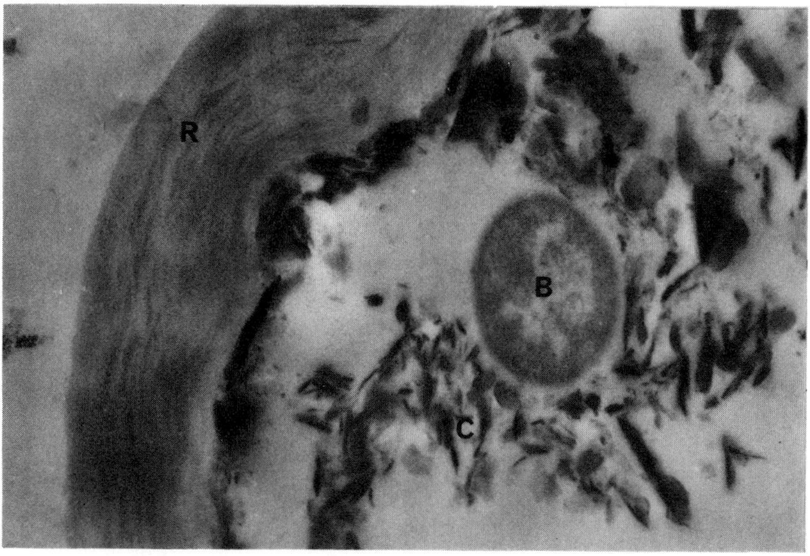

Fig. 9. A single cell of *Bacillus mycoides* enveloped by clay which is also associated with the root surface. R, root surface; C, clay particles; B, bacterial cell, × 27,700. (Photograph by J. L. Faull & R. Campbell, reproduced by permission of the National Research Council of Canada from the *Canadian Journal of Botany* (1979) **57**, 1800–1808.)

show the growth of *Bacillus mycoides* in the rhizoplane: both bacterium and root are surrounded by polysaccharide material which could contribute to the structure; the adhesion between the bacterium and the clay particles can be seen clearly. It seems that bacteria can be situated in the smallest of the soil interstices and 'cement' the smaller (clay and silt) particles together. Fungi, generally being of larger diameter, can only colonize the outside of these primary aggregates, producing a binding action (Lynch 1979). Whatever the mechanism, the action is extremely important to agriculture because although soil erosion is not as common in Britain as it is in many parts of the world, crop yields could be enhanced by improvements in soil structure. More work is therefore essential to understand the mechanisms of soil aggregate formation and stabilization and this could well lead to a re-appraisal of, for example, the benefits of grass leys to increase the microbial biomass in soils.

D. Modification of the rhizosphere by genetic manipulation of the plant

It was found by Neal *et al.* (1973) that the substitution of a pair of 5B chromosomes into a wheat cultivar changed the rhizosphere characteristics (in terms of the amount of root rot and the numbers of cellulolytic, pectinolytic, amylolytic, ammonifying and total bacteria) such that the recipient plant had a rhizosphere similar to that of the donor parent. This may have been a consequence of the different forms and amounts of carbon compounds released by the roots in the two cultivars. Dobereiner & Campelo (1971) showed that *Azotobacter paspali* established itself on the rhizoplane of tetraploid but not diploid cultivars of *Paspalum notatum*, and Gilmour & Gilmour (1978) reported that certain rice cultivars introduced before the use of inorganic N fertilizer possessed greater nitrogenase activity in their rhizospheres than in those of the cultivars then in current use. These observations should be considered in future plant breeding programmes.

E. Is inoculation a practical proposition?

Modification of the soil microbial biomass by conventional agricultural practices would be easier than attempts to modify it by means of direct applications of micro-organisms. The idea of doubling the biomass in the surface 5 cm of an average soil (*ca.* 300 kg C/ha) would involve the application of *ca.* 6 tonnes fresh weight of bacteria per hectare, which is about the normal addition of straw to soil and is of course, totally impracticable. Attempts to manipulate the microbial species balance in the rhizosphere population on a field scale have generally been by seed inoculation. This has most commonly been aimed at stimulating nitrogen-fixing populations and is discussed in the next contribution (Ch. 2, this volume). The general lack of success must cast gloom on this approach. We should however be excited by the successful attempts to control the most serious root disease of wheat, take-all (*Gaeumannomyces graminis*), either by root inoculation with *Bacillus mycoides* (Faull & Campbell 1979) or seed inoculation with fluorescent pseudomonads (R. J. Cook, pers. comm.). However, there has probably been too little consideration of the autecology of introduced organisms in relation to their synecology (Elliott *et al.* 1980). A notable exception is the studies on plant growth-promoting rhizobacteria (PGPR) which have been used as inoculants for potato, sugar beet and radish (Kloepper *et al.* 1980). A new siderophore, pseudobactin, was isolated from these bacteria and this strongly complexed the iron, making it unavailable to the plant pathogen, *Erwinia carotovora*. Pure pseudobactin had the same effect as the PGPR, and mutants which did not produce the siderophore did not promote plant growth. Production of a specific compound has also been demonstrated as the mode of action for the antagonism of *Pseudomonas fluorescens* to *Pythium ultimum* when the former was used as a seed treatment; the antifungal antibiotic, pyoluteorin, was produced by the bacterium (Howell & Stipanovic 1980).

Bowen (1980) has argued eloquently that we are only beginning to apply conventional concepts of population ecology to the rhizosphere. We must consider why we should want to change the rhizosphere biomass and species balance of the population. Plants growing under the new nutrient film conditions may well show the greatest potential for such desirable inoculation studies.

11. Conclusion

To return to the question of a possible function for the considerable loss of carbonaceous compounds from plants via their roots, let us consider the effect of this on plant productivity. Whereas micro-organisms have a variable effect on root and shoot growth (e.g. Domsch 1968; Lynch & White 1977), experiments with irradiated soil have shown that root and shoot growth were promoted in the presence of micro-organisms (Barber & Martin 1976). The reasons for this are unclear but both plant productivity and the total carbon input to soil (root and microbial biomass) were increased when micro-organisms were present around roots. This saprophytic population should therefore not be regarded as depriving the roots of energy; on the contrary it may effectively conserve energy. It would seem that one of the major functions of the associated bacteria could be to preserve in microbial form a supply of nutrients to the plant and thus prevent their loss by leaching. This would be particularly critical in soils where the availability of nutrients already limits plant productivity.

Although plants commonly grow quite well in the absence of micro-organisms in laboratory conditions, in natural conditions they are exposed to soil populations which may include antagonists of plant growth. It therefore seems that another function of the 'normal' rhizosphere microflora may be to provide a stable community that can minimize invasion by antagonists of root growth, such as pathogens, phytotoxin-producers or organisms which compete for oxygen. We must understand this 'normal' microflora better before we can hope to stabilize it by modified agricultural practices or by inoculation, with the objectives of maximizing crops and minimizing disease. For this we need to use modern autecological methods, such as chemostat studies in the laboratory, and couple them with synecological approaches in the field.

12. Summary

The rhizosphere of crop plants is described in anatomical, ecological, physiological and mathematical terms. Plants release about 20% of their total photosynthate through the roots and this provides much substrate for the microbial biomass in the rhizosphere.

Under different conditions, the rhizosphere population can promote or inhibit the uptake of ions by plants. Free-living N_2-fixing bacteria associate with roots, but rhizosphere denitrification with N loss as N_2O and N_2 has an adverse effect on the N economy of soils. Microbial products of the rhizosphere, such as phytotoxins, affect plant growth directly, but the significance of growth regulators produced by bacteria *in situ* is unclear. Other bacterial products, such as polysaccharides, influence plants indirectly by modifying soil structure. Bacteria can also affect plants adversely in conditions of poor aeration by competing for the available oxygen. These considerations are relevant to plants grown in soil, in solution and by the nutrient film technique.

A 'healthy' rhizosphere community may help to arrest the invasion of microbial antagonists of plant growth. The potentials of seed inoculation and genetic manipulation of the plant to promote these defensive measures are discussed. It is emphasized that control of the rhizosphere will be achieved only when we have a better understanding of both the autecology and synecology of bacteria growing around roots.

I thank the following for much helpful discussion and for providing photographs: Drs R. Campbell, J. L. Nicklin (*née* Faull) and E. I. Newman (University of Bristol), Professor C. M. Gilmour (University of Idaho), Dr L. F. Elliott (Washington State University), Mr M. P. Greaves and Dr J. A. Sargent (Weed Research Organization), Drs R. A. Bennett and M. C. Drew (Letcombe).

13. References

ALLAM, A. I. & HOLLIS, J. P. 1972 Sulfide inhibition of oxidases in rice roots. *Phytopathology* **62**, 634–639.

ALLEN, R. N. & NEWHOOK, F. J. 1973 Chemotaxis of zoospores of *Phytophthora cinnamomi* to ethanol in capillaries of soil pore dimensions. *Transactions of the British Mycological Society* **61**, 287–302.

ARMSTRONG, W. 1979 Aeration in higher plants. *Advances in Botanical Research* **7**, 225–232.

BALANDREAU, J. & KNOWLES, R. 1978 The rhizosphere. In *Interactions Between Non-Pathogenic Soil Micro-organisms and Plants* ed. Dommergues, Y. R. & Krupa, S. V. pp.243–268. Amsterdam: Elsevier.

BARBER, D. A. 1978 Nutrient uptake. In *Interactions Between Non-Pathogenic Soil Micro-organisms and Plants* ed. Dommergues, Y. R. & Krupa, S. V. pp.131–162. Amsterdam: Elsevier.

BARBER, D. A. & GUNN, K. B. 1974 The effect of mechanical forces on the exudation of organic substances by the roots of cereal plants grown under sterile conditions. *New Phytologist* **73**, 39–45.

BARBER, D. A. & LYNCH, J. M. 1977 Microbial growth in the rhizosphere. *Soil Biology and Biochemistry* **9**, 305–308.

BARBER, D. A. & MARTIN, J. K. 1976 The release of organic substances by cereal roots in soil. *New Phytologist* **76**, 69–80.

BARBER, D. A., BOWEN, G. D. & ROVIRA, A. D. 1976 Effects of micro-organisms on the absorption and distribution of phosphate in barley. *Australian Journal of Plant Physiology* **3**, 801–808.

BENIANS, G. J. & BARBER, D. A. 1974 The uptake of phosphate by barley plants from soil under aseptic and non-sterile conditions. *Soil Biology and Biochemistry* **6**, 195–200.

BENNETT, R. A. & LYNCH, J. M. 1981 Bacterial growth and development in the rhizosphere of gnotobiotic cereal plants. *Journal of General Microbiology* in press.

BOWEN, G. D. 1980 Misconceptions, concepts and approaches in rhizosphere biology. In *Contemporary Microbial Ecology* ed. Ellwood, D. C., Hedger, J. N., Latham, M. J., Lynch, J. M. & Slater, J. H. pp.283–304. London & New York: Academic Press.

BROWN, M. E. 1972 Plant growth substances produced by micro-organisms of soil and rhizosphere. *Journal of Applied Bacteriology* **35**, 443–451.

BROWN, M. E. 1974 Seed and root bacterization. *Annual Review of Phytopathology* **12**, 181–197.

BURFORD, J. R., DOWDELL, R. J., CREES, R. & HALL, K. C. 1979 Soil aeration and denitrification. *Agricultural Research Council Letcombe Laboratory Annual Report for 1978*, p.26.

BURNS, R. G. 1981 Microbial adhesion to soil surfaces; consequences for growth and enzyme activities. In *Microbial Adhesion to Surfaces* ed. Berkeley, R. C. W., Lynch, J. M., Melling, J., Rutter, P. R. & Vincent, B. pp.249–262. Chichester: Ellis Horwood.

CHESHIRE, M. V. 1979 *Nature and Origin of Carbohydrates in Soils.* London & New York: Academic Press.

CLARKE, A. L., GREENLAND, D. J. & QUIRK, J. P. 1967 Changes in some physical properties of the surface of an impoverished red-brown earth under pasture. *Australian Journal of Soil Research* **5**, 59–68.

COLEMAN, D. C., ANDERSON, R. V., COLE, C. V., ELLIOTT, E. T., WOODS, L. & CAMPION, M. K. 1978*a* Trophic interactions in soils as they affect energy and nutrient dynamics, IV. Flows of metabolic carbon and biomass. *Microbial Ecology* **4**, 373–380.

COLEMAN, D. C., COLE, C. V., HUNT, H. W. & KLEIN, D. A. 1978*b* Trophic interactions in soils as they affect energy and nutrient dynamics. I. Introduction. *Microbial Ecology* **4**, 345–349.

DARBYSHIRE, J. F. & GREAVES, M. P. 1973 Bacteria and protozoa in the rhizosphere. *Pesticide Science* **4**, 349–360.

DAZZO, F. B. 1981 Microbial adhesion to plant surfaces. In *Microbial Adhesion to Surfaces* ed. Berkeley, R. C. W., Lynch, J. M., Melling, J., Rutter, P. R. & Vincent, B. pp.311–328. Chichester: Ellis Horwood.

DOBEREINER, J. & CAMPELO, A. M. 1971 Non-symbiotic nitrogen-fixing bacteria in tropical soils. *Plant and Soil* Special Volume, 457–470.

DOMSCH, K. H. 1968 Microbial stimulation and inhibition of plant growth. *9th International Congress of Soil Science Transactions* **3**, 455–463.

DREW, M. C. & LYNCH, J. M. 1980 Soil anaerobiosis, micro-organisms and root function. *Annual Review of Phytopathology* **18**, 37–67.

DOMMERGUES, Y. & JACQ, V. 1972 Microbiological transformations of sulphur in the rhizosphere and spermosphere. *Annales Agronomiques* **23**, 201–215.

DUFF, R. B., WEBLEY, D. M. & SCOTT, R. O. 1963 Solubilization of minerals and related materials by 2-ketogluconic acid-producing bacteria. *Soil Science* **95**, 105–114.

ELLIOTT, L. F., GILMOUR, C. M., LYNCH, J. M. & TITTEMORE, D. 1980 Bacterial colonization of plant roots. In *Microbial-Plant Interactions* ed. Todd, R. L. Madison: Soil Science Society of America. In press.

FAULL, J. L. & CAMPBELL, R. 1979 Ultrastructure of the interaction between the take-all fungus and antagonistic bacteria. *Canadian Journal of Botany* **57**, 1800–1808.

FLETCHER, M. F., LATHAM, M. J., LYNCH, J. M. & RUTTER, P. R. 1981 Characteristics of interfaces and their rôle in microbial attachment. In *Microbial Adhesion to Surfaces* ed. Berkeley, R. C. W., Lynch, J. M., Melling, J., Rutter, P. R. & Vincent, B. pp.67–78. Chichester: Ellis Horwood.

GARCIA, J. L., RAIMBAULT, M., JACQ, V., RINAUDO, G. & ROGER, P. 1974 Activitées microbiennes dans les sols de rizières du Senegal: Relations avec les charactéristiques physico-chimiques et influence de la rhizosphère. *Revue d'Écologie et de Biologie du Sol* **11**, 169–185.

GAMBRELL, R. P. & PATRICK, W. H. 1978 Chemical and microbiological properties of anaerobic soils and sediments. In *Plant Life in Anaerobic Environments* ed. Hook, D. D. & Crawford, R. M. M. pp.375–423. Michigan: Ann Arbor.

GERRETSEN, F. C. 1948 The influence of micro-organisms on the phosphate intake by the plant. *Plant and Soil* **1**, 51–85.

GILMOUR, J. T. & GILMOUR, C. M. 1978 Nitrogenase activity of rice plant roots. *Soil Biology and Biochemistry* **10**, 261–264.

GRIFFIN, D. M. 1968 A theoretical study relating the concentration and diffusion of oxygen to the biology of organisms in soil. *New Phytologist* **67**, 561–577.

HARPER, S. H. T. & LYNCH, J. M. 1979 Effects of *Azotobacter chroococcum* on barley seed germination and seedling development. *Journal of General Microbiology* **112**, 45–51.

HARRISON, D. E. F. 1972 Physiological effects of dissolved oxygen tension and redox potential on growing populations of micro-organisms. *Journal of Applied Chemistry and Biotechnology* **22**, 417–440.

HILTNER, L. 1904 Über neuere Erfahrungen und Probleme auf dem Gebiet der Bodenbakteriologie und unter besonderer Berücksichtigung der Gründüngung und Brache. *Arbeiten der Deutschen Landwirtschaftsgesellschaft Berlin* **98**, 59–78.

HOLLIS, J. P., ALLAM, A. I., PITTS, G., JOSHI, M. M. & IBRAHIM, I. K. A. 1975 Sulfide diseases of rice on iron-excess soils. *Acta Phytopathologica Academiae Scientiarum Hungaricae* **10**, 329–341.

HOSSELL, J. C. & BAKER, J. H. 1979 Estimation of the growth rates of epiphytic bacteria and *Lemna minor* in a river. *Freshwater Biology* **9**, 319–327.

HOWELL, C. R. & STIPANOVIC, R. D. 1980 Suppression of *Pythium ultimum*-induced damping-off of cotton seedlings by *Pseudomonas fluorescens* and its antibiotic, pyoluteorin. *Phytopathology* **70**, 712–715.

JENKINSON, D. S. & POWLSON, D. S. 1976 The effects of biocidal treatments on metabolism in soil—V. A method for measuring soil biomass. *Soil Biology and Biochemistry* **8**, 209–213.

JOSHI, M. M. & HOLLIS, J. P. 1977 Interaction of *Beggiatoa* and rice plant: Detoxification of hydrogen sulfide in the rice rhizosphere. *Science, N.Y.* **195**, 179–180.

KEFFORD, N. P., BROCKWELL, J. & ZWAR, J. A. 1960 The symbiotic synthesis of auxin by legumes and nodule bacteria and its rôle in nodule development. *Australian Journal of Biological Sciences* **13**, 456–467.

KLOEPPER, J. W., LEONG, J., TEINTZE, M. & SCHROTH, M. N. 1980 Enhanced plant growth by siderophores produced by plant growth-promoting rhizobacteria. *Nature, London* **286**, 885–886.

LEE, R. B. 1980 Sources of reductant for nitrate assimilation in non-photosynthetic tissue: a review. *Plant, Cell and Environment* **3**, 65–90.

LOW, A. J. 1955 Improvements in the structural state of soils under leys. *Journal of Soil Science* **6**, 179–199.

LYNCH, J. M. 1976 Products of soil micro-organisms in relation to plant growth. *CRC Critical Reviews in Microbiology* **5**, 67–107.

LYNCH, J. M. 1979 The terrestrial environment. In *Microbial Ecology. A Conceptual Approach* ed. Lynch, J. M. & Poole, N. J. pp.67–91. Oxford: Blackwell Scientific.

LYNCH, J. M. 1981 The rhizosphere. In *Methods in Microbial Ecology* ed. Burns, R. G. & Slater, J. H. Oxford: Blackwell Scientific. In press.

LYNCH, J. M. & PANTING, L. M. 1980*a* Cultivation and the soil biomass. *Soil Biology and Biochemistry* **12**, 29–33.

LYNCH, J. M. & PANTING, L. M. 1980*b* Variations in the size of the soil biomass. *Soil Biology and Biochemistry* **12**, 547–550.

LYNCH, J. M. & WHITE, N. 1977 Effects of some non-pathogenic micro-organisms on the growth of gnotobiotic barley plants. *Plant and Soil* **47**, 161–170.

NEAL, J. L., LARSON, R. I. & ATKINSON, T. G. 1973 Changes in the rhizosphere populations of selected physiological groups of bacteria related to substitution of specific pairs of chromosomes in spring wheat. *Plant and Soil* **39**, 209–212.

NEWMAN, E. I. & BOWEN, H. J. 1974 Pattern of distribution of bacteria on root surfaces. *Soil Biology and Biochemistry* **6**, 205–209.

NEWMAN, E. I. & WATSON, A. 1977 Microbial abundance in the rhizosphere: a computer model. *Plant and Soil* **48**, 17–56.

NEWMAN, E. I., HEAP, A. J. & LAWLEY, R. A. 1981 Abundance of mycorrhizas and root-surface micro-organisms of *Plantago lanceolata* in relation to soil and vegetation: a multivariate approach. *New Phytologist* in press.

ODUNFA, V. S. A. & OSO, B. A. 1978 Bacterial population in the rhizosphere soils of cowpea and sorghum. *Revue d'Écologie et de Biologie du Sol* **15**, 413–420.

OLD, K. M. & NICOLSON, T. H. 1978 The root cortex as part of a microbial continuum. In *Microbial Ecology* ed. Loutit, M. W. & Miles, J. A. R. pp.291–294. Berlin: Springer-Verlag.

PIRT, S. J. 1975 *Principles of Microbe and Cell Cultivation*. Oxford: Blackwell Scientific.

POLONENKO, D. R., PIKE, D. J. & MAYFIELD, C. I. 1978 A method for the analysis of growth patterns of micro-organisms in soil. *Canadian Journal of Microbiology* **24**, 1262–1271.

REID, J. B. & GOSS, M. J. 1980 Changes in the aggregate stability of a sandy loam effected by growing roots of perennial ryegrass (*Lolium perenne*). *Journal of the Science of Food and Agriculture* **31**, 325–328.

ROBINSON, G. S. & JACQUES, W. A. 1958 Root development in some common New Zealand pasture plants. X. Effect of pure sowings of some grasses and clovers on the structure of a Tokomaru silt loam. *New Zealand Journal of Agricultural Research* **1**, 199–216.

ROUATT, J. W. & KATZNELSON, H. 1961 A study of the bacteria on the root surface and the rhizosphere soil of crop plants. *Journal of Applied Bacteriology* **24**, 164–171.

ROVIRA, A. D. 1965 Plant root exudates and their influence upon soil micro-organisms. In *Ecology of Soil-Borne Plant Pathogens—Prelude to Biological Control* ed. Baker, K. F. & Snyder, W. C. pp.170–186. London: John Murray.

RUSSELL, R. S. 1977 *Plant Root Systems: Their Function and Interaction with the Soil.* London: McGraw-Hill.

SAUERBECK, D. R. & JOHNEN, B. G. 1977 Root formation and decomposition during plant growth. In *Soil Organic Matter Studies* Vol. 1, pp.141–148. Vienna: International Atomic Energy Agency.

SCHIPPERS, B. & VAN VUURDE, J. W. L. 1978 Studies of microbial colonization of wheat roots and the manipulation of the rhizosphere microflora. In *Microbial Ecology* ed. Loutit, M. W. & Miles, J. A. R. pp.295–298. Berlin: Springer-Verlag.

SLANKIS, V. 1973 Hormonal relationships in mycorrhizal development. In *Ectomycorrhizae* ed. Marks, G. C. & Kozlowski, T. T. pp.231–238. London & New York: Academic Press.

SLATER, J. H. & BULL, A. T. 1978 Interactions between microbial populations. In *Companion to Microbiology* ed. Bull, A. T. & Meadow, P. M. pp.181–206. London: Longman.

SMITH, M. S. & TIEDJE, J. M. 1979 The effect of roots on soil denitrification. *Soil Science Society of America Journal* **43**, 951–955.

VANCURA, V. & KUNC, F. 1977 The effect of streptomycin and actidione on respiration in the rhizosphere and non-rhizosphere soil. *Zentralblatt für Bakteriologie, Parasitenkunde, Infektionskrankheiten und Hygiene Abt. 2* **132**, 472–478.

WILLIAMS, A. M. & WILSON, P. W. 1954 Adaptation of *Azotobacter* cells to tricarboxylic acid substrates. *Journal of Bacteriology* **67**, 353–360.

WINSOR, G. W., HURD, R. G. & PRICE, D. 1979 *Nutrient Film Technique*. Growers Bulletin No. 5. Littlehampton: Glasshouse Crops Research Institute.

Nitrogen Fixation by Free-living Bacteria Associated with Plants — Fact or Fiction?

MARGARET E. BROWN

Soil Microbiology Department, Rothamsted Experimental Station, Harpenden, Hertfordshire, UK

Contents

1. Introduction

EVER SINCE the discovery that soil contained free-living bacteria which were able to fix atmospheric nitrogen was made scientists have been interested in their role in the nitrogen economy of the soil, either in natural habitats or where the bacteria have been artificially introduced, usually in the form of so-called 'bacterial fertilizers'. The general opinion which has grown up over the years is that the bacteria fixed too little nitrogen to be of economic value, especially in countries able to use chemical fertilizers. However the recent increases in costs of these fertilizers has led to a revival in interest in biological nitrogen fixation and the use of 'bacterial fertilizers'.

2. Bacterial Fertilizers

Bacterial fertilizers were used mainly in eastern Europe and the USSR and were generally manufactured from cultures of *Azotobacter chroococcum*, the most widely distributed species of the free-living, aerobic, nitrogen-fixing bacteria. These fertilizers were said to benefit 50–70% of the field crops treated, increasing yield by 10–27%. By 1958 about 10^7 hectares in the USSR were treated with these preparations (Rubenchik 1963). However Mishustin & Naumova (1962) concluded that positive effects occurred only in about one-third of the experiments, with yield increases not usually exceeding 10%. Up to 1962 scientists from other countries had dismissed the value of these inoculants, but the appearance of the reviews by Cooper (1959) and Rubenchik (1963) renewed interest and a series of comprehensive and statistically designed experiments with different crop plants were undertaken, usually using *Azoto. chroococcum* as the inoculant. This work was reviewed by Brown (1974) who concluded that inoculation of seeds or roots led to changes in plant growth and sometimes to yield increases. The best results were usually obtained in soils with added mineral fertilizers, including nitrogen, and containing a reasonable amount of organic matter. Brown & Cooper (1963) incubated inoculated and control wheat seedlings with $^{15}N_2$ and found that rhizosphere soils from both series were slightly and similarly enriched with ^{15}N, but not the whole plants or non-rhizosphere soils. There were 10^6 Azotobacter/g of inoculated rhizosphere soil. This result, together with the fact that the inoculants worked best in soils with added N, suggested that the effects on plant growth were not the result of biological nitrogen fixation, a view that became generally accepted. There was far more evidence that the bacteria produced plant growth regulating substances in the root zone, and that these were then taken up by the plant. The evidence also indicated that in certain situations there was biological control of plant pathogens (Brown 1974).

3. Biological Nitrogen Fixation in Temperate Zones

A. The bacteria

Free-living nitrogen-fixing bacteria are present in most soils and in plant rhizospheres. There are several different bacterial species that can fix nitrogen in laboratory culture and may do so in the rhizosphere. Table 1 shows those which could fix nitrogen in temperate soils. Species of *Azotobacter* have received most attention, followed by *Clostridium* spp. and members of the Enterobacteriaceae.

TABLE 1
Bacteria that fix nitrogen in culture and may do so in temperate soils

Bacteria	General comments
Azotobacter spp.	Soil, water, leaf and root surfaces. Fix aerobically, more efficient at low O_2 tensions. Alkaline soils.
Clostridium spp.	Soil, water, sediments. Fix anaerobically.
Bacillus polymyxa	Widespread occurrence, aerobic or facultative anaerobic. Most strains fix N_2 anaerobically.
Klebsiella pneumoniae	Leaf and nodule surfaces. Fix anaerobically, or when micro-aerophilic.
Desulfovibrio spp.	Wet soils, fresh and salt water with high organic content. Few strains fix.

Data abstracted from Sprent (1979).

Natural populations of *Azotobacter* in the rhizospheres of plants are very variable in distribution and size, depending on soil, plant species and development, and fertilizer treatments. In arable soils in England non-rhizosphere populations ranged between 0 and 10^3/g soil and in rhizosphere soils up to 8 × 10^3/g. Presence of the bacteria was pH dependent; they were most abundant in neutral soils and rarely present in soils of pH less than 6·5 (Brown *et al.* 1962). Non-rhizosphere populations were also found to be low in Poland (Strelczyk 1958) and Canada (Katznelson & Strelczyk 1961). In Russia however the bacteria were often abundant, especially in organically rich soils, and they were stimulated in plant rhizospheres (Mishustin & Shilnikova 1971).

Prior to the development of sensitive methods to measure nitrogen fixation in the soil the role these different bacteria played was usually inferred from their numbers found in the ecosystem and their nitrogen fixing efficiency in pure cultures, but such inferences were not really valid.

B. Investigations using the acetylene reduction technique

The development of the sensitive acetylene reduction technique (Hardy *et al.* 1968) enabled many studies to be made on nitrogen fixation in ecosystems where reported gains in soil nitrogen could not be accounted for by non-biological sources. Table 2 shows some of the data obtained by this technique in different temperate regions.

Long-term nitrogen balance studies were made at Rothamsted in agricultural and regenerating herbaceous and woodland sites (Jenkinson 1976; Witty *et al.* 1976). The arable experiment on Broadbalk field was started in 1843, drains were installed to measure nitrogen loss through leaching, and

TABLE 2
Biological nitrogen fixation rates in temperate soils estimated from acetylene reduction measurements of soil cores with roots taken from the field

Vegetation	Location	Rate (kg N/ha/year)	Reference
Grasslands	Saskatchewan, Canada	1–2	Paul *et al.* (1971) Vlassak *et al.* (1973)
Pastures	England & Wales	4	Lockyer & Cowling (1977)
Forest	Norway, Sweden, Finland	38	Granhall & Lindberg (1976)
Woodland	England	15	Day *et al.* (1975*a*)
Woodland	England	36–109	Day *et al.* (1975*a*)
Course grassland	England	40–87	Day *et al.* (1975*a*)
Prairie	USA	0·2–0·7	Kapustka & Rice (1978)
Wheat	Broadbalk Field, England		
	No manure	7	Day *et al.* (1975*a*)
	+ NPK no Mg	10	Day *et al.* (1975*a*)
	+ farmyard manure	6·2	Day *et al.* (1975*a*)

soil samples have been taken at intervals since 1865. The outstanding feature of the experiment was the ability of the soil to support consecutive wheat crops without addition of nitrogenous fertilizer. The area which received fertilizer P, K and Mg, but no N, gave a mean grain yield of 2·4 t/ha over the period 1970–1972. This was about half the yield from the plot receiving optimal amounts of N. The nitrogen demand for the consecutive crops was probably met through non-biological sources and through biological nitrogen fixation; the two together might have contributed about 36 kg N/ha/year. Blue-green algae (cyanobacteria) probably in surface crusts produced most of the fixed nitrogen, but there was some evidence that nitrogen fixing bacteria in soil and rhizosphere gave *ca.* 2–3 kg N/ha/year (Day *et al.* 1975*a*). Part of this continuous wheat experiment was fenced off in 1883 and the natural vegetation allowed to re-establish to give an area of mixed deciduous woodland (Broadbalk Wilderness), and in a part of this area tree seedlings were removed annually (stubbed), leaving the ground covered by mixed herbaceous vegetation. Both sections have gained nitrogen steadily since 1883, calculated at 49 kg N/ha/year in the top 23 cm of soil in the stubbed part (Jenkinson 1971). Biological nitrogen fixation was estimated to contribute about 34 kg N/ha/year. Nitrogenase activity was detected in the rhizospheres of the herbaceous dicotyledonous weeds common to the site (Harris & Dart 1973). In the woodland part biological fixation possibly contributed about 15 kg N/ha/year (Day *et al.*

1975*a*). When conditions were wet a single sampled core with roots fixed *ca.* 0·5 kg N/ha/day, indicating a very high potential for fixation. Further investigations into the effects of soil moisture in these sites showed a highly significant and positive correlation. Isolations from the rhizospheres of the weeds suggested that *Enterobacter–Klebsiella* type bacteria were the most likely nitrogen fixing organisms.

Paul *et al.* (1971) and Vlassak *et al.* (1973) in Canada reported much lower fixation rates. This may have been due partly to different soil temperatures or to differences in the state of equilibrium of the vegetation in this system compared with Broadbalk Wilderness (Day *et al.* 1975*a*). The prairie grasslands having been undisturbed for centuries might have reached a state where the soil nitrogen content was sufficiently high and nitrogen was efficiently cycled within the ecosystem.

4. Biological Nitrogen Fixation in Tropical or Subtropical Areas

A. Associations between bacteria and plants

Over the last 10 years biological nitrogen fixation in tropical or subtropical soils has received much attention. Various associations between roots and nitrogen fixing bacteria have been described, e.g. sugar cane and other tropical grasses with *Beijerinckia* spp. (Dobereiner 1961), *Paspalum notatum* var. *batatais* with *Azotobacter paspali* (Dobereiner 1966), many tropical forage grasses with *Azospirillum brasilense* (formerly *Spirillum lipoferum*) (Day *et al.* 1975*b*) and rice with *Beijerinckia* sp. (Becking 1976). These bacteria were abundant in the rhizospheres of the different plants and *Azoto. paspali* and *Azosp. brasilense* seemed to form semi-symbiotic associations with their hosts, colonizing the mucigel, intercellular spaces and some cortical cells of the roots and rhizomes (Dobereiner & Campelo 1971, Dobereiner *et al.* 1972, Dobereiner & Day 1976). Patriquin & Dobereiner (1978), using the vital stain triphenyltetrazolium chloride, demonstrated that even the inner cortex and stele of maize roots were colonized by *Azosp. brasilense*. The bacteria remained viable even after treating the roots with sterilizing agents, indicating that the endodermis was intact. The authors presumed that these infections occurred initially in branch roots and then spread into the main roots, because high nitrogenase activity was often detected in highly branched roots with an intact cortex. These spirilla were also found in adjacent non-rhizosphere soil.

Such a close affinity between host and bacteria suggested there might be mutual benefit. Dobereiner *et al.* (1972) first studied the *Azotobacter–Paspalum* association using the acetylene reduction assay and found nitrogenase activity in soil cores with intact plants and on washed roots and rhizomes, which, when interpreted as nitrogen fixed per hectare per year, amounted to values up to 93 kg for cores, and 117 kg for roots. The nitrogenase activity was detected after a lag phase of *ca.* 12 h. It was thought that this amount of fixation would be adequate for the growth requirements of *P. notatum* in the field and was consistent with the plant's ability to establish an extensive cover on poor soils without any nitrogen fertilizer. Acetylene reduction assays on roots of maize and of *Digitaria decumbens* associated with *Azosp. brasilense* indicated even greater amounts of nitrogen fixation, equivalent to an input of 876 kg N/ha/year (von Bülow & Dobereiner 1975, Dobereiner & Day 1976). Naturally, reports of biological fixation of this order aroused considerable interest and stimulated further research into the associations.

B. Nitrogen fixation in Azotobacter paspali *associations*

Azotobacter paspali was associated only with certain varieties of *Paspalum notatum*, notably the tetraploid cultivar *batatais*. The fine leaved diploid cultivars such as *pensacola* were unable to support the bacteria, and ethylene production by root samples from these diploids was never more than 29 nmol/g/h and at some seasons of the year could not be detected (Day *et al.* 1975*b*). Attempts by Kass *et al.* (1971) to inoculate *Azoto. paspali* into the rhizospheres of *P. notatum* grown in sand culture were only successful with the *batatais* cultivar, and then only when glucose was added at the time of inoculation. There was no evidence of improved growth or for nitrogen gains in the leaves but nitrogen gains by each root system were equivalent to an 18% increase in the nitrogen of the root mass. These high levels of nitrogen in the roots might have resulted from an accumulation of *Azoto. paspali* on the root surface rather than uptake of nitrogen fixed. They also found that soils with *P. notatum* and *Azoto. paspali* showed some nitrogen fixation. These gains might have resulted from fixation by photosynthetic bacteria which were present in the soil. When bare soil was incubated in the dark all nitrogen fixation ceased.

Work by Dobereiner *et al.* (1972), Balandreau *et al.* (1971) and Dommergues *et al.* (1972) strongly suggested a relationship between photosynthesis and nitrogenase activity. Dobereiner *et al.* (1972) found that nitrogenase activity almost ceased if soil cores with *P. notatum* were kept dark for 60 h, but rapidly returned if the cores were put back in the light. If the cores were given a 16 h day regime acetylene reduction was linear and unaffected by the diurnal rhythm. However, van Berkum (1978) found that field-grown plants of *Secale cereale*, *Secale vulgare* and *Brachiaria mutica* all showed diurnal variation in nitrogenase activity; minimal activity occurred between 00.00 h and 06.00 h.

C. Nitrogen fixation in Azospirillum brasilense *associations*

The *Azospirillum* association with many different forage grasses was considered to be of much greater importance than the *Azotobacter–Paspalum* association but results from many further investigations of natural ecosystems and of inoculated plants have proved disappointing. Tjepkema *et al.* (1976) failed to obtain a significant activity from more than 100 cores of several non-legumes, although the cores contained active bacteria, nor were they any more successful with sorghum and maize breeding lines; fixation rates never exceeded 2 g N/ha/day and the average dry weight of plants was *ca.* 60% less than that of those receiving NH_4NO_3. They did obtain fixation by excised pre-incubated roots. As the soils were dry and sandy they concluded that better fixation might occur in wetter conditions. Later they inoculated pearl millet (*Panicum maximum*) grown in well-watered pots of soil but did not obtain fixation rates of more than 23 g N/ha/day, either from intact systems or roots of plants (Barber *et al.* 1979). Burris *et al.* (1976) in axenic maize cultures inoculated with *Azosp. brasilense*, obtained fixation rates which represented the addition of 1 kg N/ha for the months of the growing season, whilst cores from soil grown maize gave rates of < 2 kg N/ha/season, but excised roots reached 10 kg N/ha/season. Smith *et al.* (1976) were more successful when they tested 40 genotypes representing five tropical grasses. Yields from *Digitaria decumbens* and *Panicum maximum* were increased by 163% and 15%, respectively, following inoculation with *Azosp. brasilense*; *P. maximum* also had a higher content of crude protein. Adding nitrogen fertilizer at rates of 40 and 80 kg/ha to inoculated plots of pearl millet significantly increased yields over uninoculated controls. Had only fertilizer been used it would have been necessary to add 60 and 122 kg N/ha, respectively, to obtain equivalent yields. Taylor (1979) obtained similar results with 89 kg/ha of fertilizer N; 134 kg N/ha did not have the additive effect with the bacteria. Measurements of nitrogenase activity were not given by these authors and the increases in yields might have arisen from plant responses to growth regulator production by the bacteria.

D. Possible role of plant growth regulators

Barea & Brown (1974) inoculated a variety of dicotyledons and monocotyledons growing in pots in natural soil with *Azotobacter paspali* and obtained large increases in early plant growth. For example, the stem length of tomatoes with four fully developed leaves was increased by 145% compared with controls, but these differences had disappeared when 12 leaves had formed. Two weeks after treatment the number of leaves of *Paspalum notatum* had increased by 18% and their length by 49% and dry weights were significantly greater than controls. These effects were maintained for four weeks before decreasing. Nitrogenase activity could not be detected on whole unwashed root systems from inoculated or control plants, although *Azoto. paspali* cells were plentiful on the inoculated roots. In later experiments Brown (1976) could not detect nitrogenase activity on roots of inoculated young plants growing in soil but did detect it on roots of 14-week-old plants. Activity was similar whether plants had been inoculated with living *Azoto. paspali*, sterile culture supernatant fluids or diluted culture medium without sucrose. The culture fluids contained at least three types of plant growth regulator, indolyl 3-acetic acid (IAA), gibberellins and cytokinins, in amounts that were previously shown to influence the growth of young *Paspalum* plants (Barea & Brown 1974). The authors suggested that nitrogen fixation by *Azoto. paspali* might have been of secondary importance to growth regulator production, especially when soil nitrogen was adequate for plant growth.

Gaskins & Hubbell (1979) found that growth of pearl millet seedlings in soil was improved by inoculation with living or dead cells of *Azospirillum*, whilst acetylene reduction rates of intact root systems were not increased. When gibberellic acid, kinetin or IAA was added to the soil the short term growth responses were similar to those of adding the bacteria. Plant growth regulators on chromatograms of cultures of *Azosp. brasilense* were identified, tentatively, as IAA and cytokinin. Reynders & Vlassak (1979) found that IAA was formed in cultures of *Azosp. brasilense* after adding tryptophan to the medium and they suggested that auxins were responsible for the increased growth of plants with many *Azosp. brasilense* cells on or inside their roots.

5. The Acetylene Reduction Technique for Assessing Nitrogenase Activity

The method is an indirect one based on the fact that the nitrogenase enzyme reduces acetylene to ethylene which can be assayed at very low concentrations by gas chromatography. With systems such as legume root nodules the ethylene appears immediately following the addition of the acetylene and is produced at a constant rate over several hours, but with rhizosphere fixation, measured either on washed roots or from intact soil cores, there is a lag period following the addition of the acetylene and production of ethylene is seldom linear.

A. Nitrogenase activity associated with excised washed roots

The technique for measuring nitrogenase activity of excised washed roots was developed to study the close association between roots and bacteria, because activity in soil cores might represent fixation by non-associated free-living micro-organisms, or be due to endogenous ethylene evolution by the roots or soil micro-organisms. Pre-incubation of soil cores, especially if they had been watered, could give erroneously high estimates of nitrogen fixation, as watering could cause anaerobiosis, and thus promote nitrogenase activity by anaerobic and facultative anaerobic bacteria and allow time for them to multiply. Van Berkum (1978) found that nitrogenase activity of pre-incubated, excised roots of plants associated with *Azoto. paspali* was many times greater than the activities of the corresponding soil cores, and he suggested that measurements made on soil cores taken immediately on collection approximated more closely to the rates of fixation occurring in the field. His investigation of the excised, washed root technique showed that the lag period of 8–18 h before the onset of acetylene reduction was not due to lack of nutritional requirements for nitrogen fixation, nor to damage of the nitrogenase by oxygen. The development of detectable nitrogenase activity during the pre-incubation period apparently depended on the multiplication of nitrogen fixing organisms. If combined nitrogen was added to these roots the lag phase was extended and he suggested that the nitrogen fixing bacteria used the combined nitrogen for growth until the creation of nitrogen deficiency favoured the synthesis of nitrogenase. Unwashed roots of *Secale vulgare*, *Zea mays* and *P. notatum* developed little nitrogenase activity during the pre-incubation phase suggesting that the nitrogen fixing bacteria required water for multiplication, probably by promoting the movement of substrates from the roots. Any oxygen in the water would be removed rapidly by bacterial metabolism thus creating sites with reduced oxygen tension that favoured fermentation. The excised root technique was also criticized by Barber *et al.* (1979) and Okon *et al.* (1977) after investigating activity of *Azospirillum*. They concluded that the bacteria multiplied and fixed nitrogen during pre-incubation, which led to artificially high estimates of fixation. It is now generally agreed that rates of nitrogenase activity measured from washed, excised roots bear little relationship to those occurring naturally and the procedure is unsuitable for evaluating rhizosphere-associated nitrogen fixation.

B. Nitrogenase activity associated with soil cores

When assaying soil cores Day *et al.* (1975*a*) thought that the lag period was probably caused by slow diffusion of acetylene and of ethylene through the core. Witty (1979) using labelled acetylene found that values for labelled ethylene produced from nodules of *Phaseolus vulgaris*, from *Anabaena* *cylindrica* and from anaerobic suspensions of *Bacillus polymyxa* and *Clostridum pasteurianum* fitted well to the predicted values.

The predicted values were based on $^{14}C_2H_2$ calibration curves and represented the quantity of label which should occur in the C_2H_4 if it was derived entirely from C_2H_2. Labelled ethylene produced from soil cores containing relatively high activity also approximated to the theoretical amount, but with cores of low activity the produced $^{14}C_2H_4$ was much less than the predicted amount. Thus in temperate regions, where fixation is low, errors could become significant, especially when C_2H_2-dependent C_2H_4 production extrapolated to less than 100 g N/ha/day. Witty (1979) recommended that data from soil systems fixing less than this should be interpreted with caution unless $^{14}C_2H_2$ controls were used. Van Berkum (1978) showed that ethylene in soil cores with tropical grasses might also partly be due to endogenous evolution by roots or soil micro-organisms.

Nitrogen fixation in the plant rhizosphere was confirmed by $^{15}N_2$ studies, which also gave evidence of direct and apparently quite rapid transfer of some of the fixed nitrogen to the plant (Ruschel *et al.* 1975; De Polli *et al.* 1977).

6. The Rhizosphere as a Site for Nitrogen Fixation

The rhizosphere is one region in soil where there is a constant supply of substrates readily available to the heterotrophic microflora, including nitrogen fixers, and bacterial multiplication is rapid. This leads to depletion of substrates and reduction in oxygen tensions. The nitrogen fixers only switch to fixation when nitrogen substrate and oxygen are depleted but they still require plenty of carbohydrate. Provided this is available the fixers have a competitive advantage over other organisms.

A. Substrates available in soil and rhizosphere

Nutrients leaking from roots and cell debris are the two main sources of substrate. It has been shown that the sum of exudates and deposits corresponded to *ca.* 15–25% of the weight increments of seedling wheat and barley (Barber & Martin 1976) and to *ca.* 300% of the root dry weight of mature wheat plants (von Sauerbeck *et al.* 1976). Griffin *et al.* (1976) estimated that *ca.* 0·15% of root carbon, nitrogen and hydrogen were lost per week. In the rhizosphere this material would stay very close to the roots, acting as substrate. The majority of the nitrogen fixers cannot use cellulose, hemicellulose or lignin and therefore depend on their degradation by other microbes to more acceptable substrates. The breakdown substrates appear to be very suitable as energy sources because Barrow & Jenkinson (1962), Delwiche & Wijler (1956) and Hüser (1970) all showed that if straw, dead roots or cellulose were incorporated into soil and the mixture incubated at 25°C measurable fixation followed under aerobic or anaerobic conditions, and especially under conditions of water saturation. It was suggested that with wheat straw the degradation of hemicellulose, followed by cellulose, supported successive peaks of nitrogen fixation for which organisms similar to *Cl. butyricum* appeared to be responsible. In field conditions in temperate regions fixation is likely to be less than that measured in these experiments because the soil temperatures of 10–15°C are less favourable for microbial activity. Rice *et al.* (1967) thought the aerobic–anaerobic interface in soil crumbs was very important and the degradation products of hemicellulose and cellulose in a 2 mm thick aerobic zone would diffuse across the interface to support fixation by clostridia in the anaerobic zone. Magdoff & Bouldin (1970) using an aerobic fixer (probably *Azotobacter*) in a soil–straw mixture claimed that products of anaerobic cellulose decomposition diffused into the aerobic zone to support fixation by the aerobe.

The natural substrates for soil nitrogen fixing bacteria are likely to be ethanol, butanol, phenols, acetate, butyrate and similar products arising from incomplete degradation of plant residues and from anaerobic fermentation by bacteria such as *Cl. butyricum*. Generally these substrates do not support such rapid growth as sugars (LaRue 1976), but sugars may also be found in plant root exudates (Rovira 1969).

Many of the tropical plants supporting significant fixation in the rhizosphere are those with the C_4-dicarboxylic acid pathway of photosynthesis. They form more photosynthates and root excretions than plants with the C_3 pathway (Ehleringer 1978) and take advantage of the higher temperatures and light intensities of their environment. Malate, a metabolite of the C_4 pathway is readily used by *Azosp. brasilense* and *Azoto. paspali*, and Day & Dobereiner (1976) showed that in culture under reduced oxygen tension the efficiency of carbohydrate utilization by these organisms was greater when the carbohydrate was supplied gradually rather than all at once. Gradual supply is likely to occur in the rhizosphere.

Azospirillum and *Azotobacter* species, when fixing nitrogen under substrate limitation, are always very sensitive to ambient oxygen concentrations. In the field, oxygen diffusion to the rhizosphere is limited by the extent of water saturation of the pore spaces and significant correlations have been found between rainfall, soil moisture, temperature and the *in situ* nitrogenase activity of the rhizosphere (Balandreau & Dommergues 1973; Balandreau *et al.* 1975, 1976).

The nitrogen substrates supplied by root exudates and from combined nitrogen sources affect nitrogenase activity. Any source readily yielding ammonia is generally repressive, but small amounts of ammonia or amino acids may stimulate growth of the nitrogen fixers and fixation. They have no effect on the final amount of nitrogen fixed but serve only to 'prime' the fermentation in nitrogen-free medium (Winogradsky 1895). Amino acids such as aspartic acid, asparagine and glutamic acid which are present in larger quantities in root exudates than other amino acids (Rovira 1969) were found to be the only ones supporting growth of *Azoto. chroococcum* (LaRue 1976).

B. Effects of fertilizers

Although the rhizosphere apparently offers the best environment for fixation, in practice the amounts fixed there are small. However, manipulation of the environment with fertilizer treatments might lead to increased efficiency of the system. For example, Balandreau & Dommergues (1973) showed a three-fold increase in nitrogenase activity of maize fertilized with P alone, compared with plants given N and P. Smith *et al.* (1976) and Taylor (1979) found that 40 and 80 kg N/ha added to millet inoculated with *Azosp. brasilense* improved the effect of the inoculant on crop yield. This suggested that when *Azospirillum* was present, less fertilizer N was required and inoculation could thus lead to a saving of chemical fertilizer.

7. Genetic Manipulation to Improve Nitrogen Fixation

The overall impression from experiments designed to enhance 'associative symbioses' by inoculation with the requisite bacteria is not one of great success. However, there are other ways in which the 'associative symbioses' might be made to function more efficiently, such as by genetic manipulation of host or bacterium. Breeding from plants which naturally associate with the bacteria to produce lines that provide a better habitat in which the bacteria can multiply and fix is possible. The converse has been done to the detriment of fixation, when breeding from tetraploid *Paspalum* species to produce better forage swards of diploid varieties led to the elimination of the 'associative symbiosis' (Day *et al.* 1975*b*). Alternatively, bacteria could be bred that fixed nitrogen more efficiently, e.g. bacteria requiring less carbon substrate for each molecule of nitrogen produced would make less demand on the carbon compounds available in the rhizosphere. The problem would still remain of introducing these bacteria into the rhizosphere where they would be competing with their less efficient parent strains.

The close associations between plant and bacteria do not appear to be confined to the tropical environment and breeding should also be considered for temperate plants. Neal & Larson (1976) isolated a bacillus possessing nitrogenase activity from the rhizosphere of a chromosome-substitution line of spring wheat. The bacillus was not isolated from the parent cultivar nor from a corresponding line involving an homologous chromosome. Shearman *et al.* (1979) found that only a 'Park' turf of Kentucky blue-grass (*Poa pratensis*) inoculated with a particular strain of *Klebsiella pneumoniae* showed a significant associative effect. These authors are investigating the potential for associative fixation among 15 lines of 'Park' Kentucky blue-grass.

8. General Conclusions

Nitrogen fixation in the rhizosphere is a fact, but the evidence now available suggests that the high values of nitrogen fixed per hectare per year estimated from excised, washed roots are somewhat exaggerated. However, there may be situations when fixation makes a very positive contribution to the nitrogen of the ecosystem. The requirements for this are a plentiful supply of carbon substrate, low amounts of available N, low oxygen tensions and a reasonable level of soil moisture. Such requirements are sometimes met in the rhizosphere of both temperate and tropical plants.

The extent and significance of fixation varies with geographical location and although most recently attention has been focussed on the grass associations in the tropics and subtropics, it should not be forgotten that there may be significant contributions in other regions. In polar regions and areas where peats predominate, carbon substrates are plentiful and aerobes and anaerobes probably contribute at least as much nitrogen as does precipitation (Granhall & Torsvik 1975).

Nitrogen fixation may also be important in forest ecosystems, estimates of up to 38 kg N/ha/year have been given (Granhall & Lindberg 1976; Todd *et al.* 1976). Although much of this may come from the activity of blue-green algae, both aerobic and anaerobic bacteria contribute in rhizospheres or in decaying woody litter.

Genetic manipulation of either host or bacteria forming the 'associative symbioses' may offer the best chance of obtaining systems where the nitrogen fixed is of economic importance. However before such long-term breeding trials are undertaken it is essential that the assessment of nitrogenase activity of the test association is accurate and the use of $^{15}N_2$ may be absolutely necessary. Extrapolation from values of nitrogenase activity obtained from short-term laboratory experiments with the acetylene reduction technique can only be regarded as indicative of the potential.

Although free-living bacteria at the moment are unlikely to make much impact in agriculture because they cannot produce a significant proportion of the fixed nitrogen that a growing crop demands, they are probably a vital part of all natural ecosystems. In nature it is the gradual accumulation of fixed nitrogen that matters.

9. References

BALANDREAU, J., WEINHARD, P., RINDAUDO, G. & DOMMERGUES, Y. 1971 Influence de l'intensité de l'éclairement de la plante sur la fixation non symbiotique de l'azote dans le rhizosphère. *Oecologia Plantarum* **6**, 341–351.

BALANDREAU, J. & DOMMERGUES, Y. 1973 Assaying nitrogenase (C_2H_2) activity in the field. In *Modern Methods in the Study of Microbial Ecology* ed. Rosswall, T. pp.247–254. *Symposium, Uppsala 1972.* Bulletin No. 17 from the Ecological Research Committee. Stockholm: Swedish Natural Science Research Council (NFR).

BALANDREAU, J., RINAUDO, G., IBTISSAM, F. & DOMMERGUES, Y. 1975 Nitrogen fixation in the rhizosphere of rice plants. In *Nitrogen Fixation by Free-living Micro-organisms.* International Biological Programme Vol. 6, ed. Stewart, W. D. P. pp.57–70. Cambridge: Cambridge University Press.

BALANDREAU, J., MILLIER, C., WEINHARD, P., DUCERF, P. & DOMMERGUES, Y. 1976 Études des variations de la function d'azote dans une culture de mais. *Compte rendu des séances de l'Académie des Sciences* Ser D. **282**, 1071–1074.

BARBER, D. A. & MARTIN, J. K. 1976 Release of organic substances by cereal roots into soil. *New Phytologist* **76**, 69–80.

BARBER, L. E., RUSSELL, S. A. & EVANS, H. J. 1979 Inoculation of millet with *Azospirillum. Plant and Soil* **52**, 49–57.

BAREA, J. M. & BROWN, M. E. 1974 Effects on plant growth produced by *Azotobacter paspali* related to synthesis of plant growth regulating substances. *Journal of Applied Bacteriology* **37**, 583–593.

BARROW, N. J. & JENKINSON, D. S. 1962 The effect of water-logging on fixation of nitrogen by soil incubated with straw. *Plant and Soil* **16**, 258–262.

BECKING, J. H. 1976 *Beijerinckia* in irrigated rice soils. In *Environmental Roles of Nitrogen Fixing Blue-green Algae and Asymbiotic Bacteria* ed. Granhall, U. pp.116–129. Bulletin No. 26 from the Ecological Research Committee. Stockholm: Swedish Natural Science Research Council (NFR).

BROWN, M. E. 1974 Seed and root bacterization. *Annual Review of Phytopathology* **12**, 181–197.

BROWN, M. E. 1976 Role of *Azotobacter paspali* in association with *Paspalum notatum. Journal of Applied Bacteriology* **40**, 341–348.

BROWN, M. E. & COOPER, R. 1963 Pot experiments on *Azotobacter* inoculation. *Rothamsted Report for 1962* Part 1, p.82.

BROWN, M. E., BURLINGHAM, S. K. & JACKSON, R. M. 1962 Studies on *Azotobacter* species in soil. II. Populations of *Azotobacter* in the rhizosphere and effects of artificial inoculation. *Plant and Soil* **17**, 320–332.

BURRIS, R. H., OKON, Y. & ALBRECHT, S. L. 1976 Properties and reactions of *Spirillum lipoferum*. In *Environmental Role of Nitrogen Fixing Blue-green Algae and Asymbiotic Bacteria* ed. Granhall, U. pp.353–363. Bulletin No. 26 from the Ecological Research Committee. Stockholm: Swedish Natural Science Research Council (NFR).

COOPER R. 1959 Bacterial fertilizers in the Soviet Union. *Soils and Fertilizers* **22**, 327–333.

DAY, J. M. & DOBEREINER, J. 1976 Physiological aspects of N_2-fixation by *Spirillum* from *Digitaria* roots. *Soil Biology and Biochemistry* **8**, 45–50.

DAY, J. M., HARRIS, D., DART, P. J. & VAN BERKUM, P. 1975*a* The Broadbalk experiment. An investigation of nitrogen gains from non-symbiotic nitrogen fixation. In *Nitrogen Fixation by Free-living Micro-organisms*. International Biological Programme Vol. 6, ed. Stewart, W. D. P. pp.71–84. Cambridge: Cambridge University Press.

DAY, J. M., NEVES, M. C. P. & DOBEREINER, J. 1975*b* Nitrogenase activity on the roots of tropical forage grasses. *Soil Biology and Biochemistry* **7**, 107–112.

DELWICHE, C. C. & WIJLER, J. 1956 Non-symbiotic fixation in soil. *Plant and Soil* **7**, 113–129.

DE POLLI, H., MATSUI, E., DOBEREINER, J. & SALATI, E. 1977 Confirmation of nitrogen fixation in two tropical grasses by $^{15}N_2$ incorporation. *Soil Biology and Biochemistry* **9**, 119–123.

DOBEREINER, J. 1961 Nitrogen-fixing bacteria of the genus *Beijerinckia* Derx. in the rhizosphere of sugar cane. *Plant and Soil* **15**, 211–217.

DOBEREINER, J. 1966 *Azotobacter paspali* sp. n. una bacteria fixadora de nitrogênio na rizosfera de *Paspalum. Pesquisa Agropecúária, Brasil* **1**, 357–365.

DOBEREINER, J. & CAMPELO, A. B. 1971 Non-symbiotic nitrogen fixing bacteria in tropical soils. *Plant and Soil* Special Volume, 457–470.

DOBEREINER, J. & DAY, J. M. 1976 Associative symbioses in tropical grasses. Characterization of micro-organisms and dinitrogen-fixing sites. In *Symposium on Nitrogen Fixation* Vol. 2, ed. Newman, W. E. & Newman, C. J. pp.516–538. Pullman: Washington State University Press.

DOBEREINER, J., DAY, J. M. & DART, P. J. 1972 Nitrogenase activity and oxygen sensitivity of the *Paspalum notatum-Azotobacter paspali* association. *Journal of General Microbiology* **71**, 103–116.

DOMMERGUES, Y., BALANDREAU, J., RINAUDO, G. & WEINHARD, P. 1972 Non-symbiotic nitrogen fixation in the rhizosphere of rice, maize and different tropical grasses. *Soil Biology and Biochemistry* **5**, 83–89.

EHLERINGER, J. R. 1978 Implications of quantum yield differences on the distribution of C_3 and C_4 grasses. *Oekologia* **31**, 255–267.

GASKINS, M. H. & HUBBELL, D. H. 1979 Response of non-leguminous plants to root inoculation with free-living diazotrophic bacteria. In *The Soil-Root Interface* ed. Harley, J. L. & Scott Russell, R. pp.176–182. London & New York: Academic Press.

GRANHALL, U. & LINDBERG, T. 1976 Nitrogen fixation in some coniferous forest ecosystems. In *Environmental Role of Nitrogen Fixing Blue-green Algae and Asymbiotic Bacteria* ed. Granhall, U. pp.178–192. Bulletin No. 26 from the Ecological Research Committee. Stockholm: Swedish Natural Science Research Council (NFR).

GRANHALL, U. & TORSVIK, V. 1975 Nitrogen fixation by bacteria and free-living blue-green algae in tundra areas. In *Fennoscandian Tundra Ecosystems* Part 1, ed. Wielgolaski, F. E. Berlin, Heidelberg, New York: Springer-Verlag.

GRIFFIN, G. J., HALE, M. G. & SHAY, F. J. 1976 Nature and quantity of sloughed organic matter produced by roots of axenic peanut plants. *Soil Biology and Biochemistry* **8**, 20–32.

HARDY, R. W. F., HOLSTEN, R. D., JACKSON, E. K. & BURNS, R. C. 1968 The acetylene-ethylene assay for N_2 fixation; laboratory and field evaluation. *Plant Physiology* **43**, 1185–1207.

HARRIS, D. & DART, P. J. 1973 Nitrogenase activity in the rhizosphere of *Stachys sylvatica* and some other dicotyledonous plants. *Soil Biology and Biochemistry* **5**, 277–279.

HÜSER, R. 1970 Die Abhängigkeit der Luftstickstoffbindung in Buchenmullproben von der Dosierung des Cellulosezusatzes. *Plant and Soil* **33**, 727–728.

JENKINSON, D. S. 1971 The accumulation of organic matter in soil left uncultivated. *Rothamsted Report for 1970* Part 2, pp.113–137.

JENKINSON, D. S. 1976 The nitrogen economy of the Broadbalk Experiments 1. Nitrogen balance in the experiments. *Rothamsted Report for 1975* Part 2, pp.103–109.

KAPUSTKA, L. A. & RICE, E. L. 1978 Symbiotic and asymbiotic N_2 fixation in a tall grass prairie. *Soil Biology and Biochemistry* **10**, 553–554.

KASS, D. L., DROSDOFF, M. & ALEXANDER, M. 1971 Nitrogen fixation by *Azotobacter paspali* in association with Bahia grass (*Paspalum notatum*). *Proceedings of the Soil Science Society of America* **35**, 286–289.

KATZNELSON, H. & STRELCZYK, E. 1961 Studies on the interaction of plants and free-living nitrogen fixing micro-organisms. 1. Occurrence of *Azotobacter* in the rhizosphere of crop plants. *Canadian Journal of Microbiology* **7**, 437–446.

LARUE, T. A. 1976 The Bacteria. In *A Treatise on Dinitrogen Fixation* Vol. 3, ed. Hardy, R. W. F. & Silver, W. S. pp.19–62. New York: John Wiley.

LOCKYER, D. R. & COWLING, D. W. 1977 Non-symbiotic nitrogen fixation in some soils in England and Wales. *Journal of The British Grassland Society* **32**, 7–12.

MAGDOFF, F. R. & BOULDIN, D. R. 1970 Nitrogen fixation in submerged soil-sand-energy material media and the aerobic-anaerobic interface. *Plant and Soil* **33**, 49–61.

MISHUSTIN, E. N. & NAUMOVA, A. N. 1962 Bacteria fertilizers their effectiveness and mode of action. *Mikrobiologiya* **31**, 543–555.

MISHUSTIN, E. N. & SHILNIKOVA, V. K. 1971 Structure and development of *Azotobacter*. In *Biological Fixation of Atmospheric Nitrogen* pp.184–250. London: Macmillan.

NEAL, J. L. & LARSON, R. L. 1976 Acetylene reduction by bacteria isolated from the rhizosphere of wheat. *Soil Biology and Biochemistry* **8**, 151–155.

OKON, Y., ALBRECHT, S. L. & BURRIS, R. H. 1977 Methods for growing *Spirillum lipoferum* and for counting it in pure culture and in association with plants. *Applied and Environmental Microbiology* **33**, 85–88.

PATRIQUIN, D. G. & DOBEREINER, J. 1978 Light microscopic observations of tetrazolium reducing bacteria in the endorhizosphere of maize and other grasses in Brazil. *Canadian Journal of Microbiology* **24**, 734–742.

PAUL, E. A., MYERS, R. J. K. & RICE, W. A. 1971 Nitrogen fixation in grassland and associated cultivated ecosystems. *Plant and Soil* Special volume, 495–507.

REYNDERS, L. & VLASSAK, K. 1979 Conversion of tryptophan to indole acetic acid by *Azospirillum brasilense*. *Soil Biology and Biochemistry* **11**, 547–548.

RICE, W. A., PAUL, E. A . & WETTER, L. R. 1967 The role of anaerobiosis in asymbiotic nitrogen fixation. *Canadian Journal of Microbiology* **13**, 829–836.

ROVIRA, A. D. 1969 Plant root exudates. *Botanical Reviews* **35**, 35–57.

RUBENCHIK, L. I. 1963 *Azotobacter and Its Use in Agriculture*. Jerusalem: Israel Program for Scientific Translations.

RUSCHEL, A. P., HENIS, Y. & SALATI, E. 1975 Nitrogen-15 tracing of N fixation with soil-grown sugar cane seedlings. *Soil Biology and Biochemistry* **7**, 181–182.

SHEARMAN, R. C., PEDERSEN, W. L., KLUCAS, R. V. & KINBACHER, E. J. 1979 Nitrogen fixation associated with 'Park' Kentucky blue grass (*Poa pratensis* L). *Canadian Journal of Microbiology* **25**, 1197–1200.

SMITH, R. L., BOUTON, J. H., SCHANK, S. C., QUESENBERG, K. H., TYLER, M. E., MILAM, J. R., GASKINS, M. H. & LITTELL, R. C. 1976 Nitrogen fixation in grasses inoculated with *Spirillum lipoferum*. *Science, N. Y.* **193**, 1003–1005.

SPRENT, J. L. 1979 *The Biology of Nitrogen-fixing Organisms*. Maidenhead, UK: McGraw-Hill.

STRELCZYK, E. 1958 The effect of various cultivated plants on the number of *Azotobacter* and *Clostridium* in their rhizospheres. *Acta microbiologica polonica* **7**, 115–123.

TAYLOR, R. W. 1979 Response of 2 grasses to inoculation with *Azospirillum* spp. in a Bahamian soil. *Tropical Agriculture* **56**, 361–365.

TJEPKEMA, J. D., RUSSELL, S. A. & EVANS, H. J. 1976 Acetylene reduction (nitrogen fixation) associated with corn inoculated with *Spirillum*. *Applied and Environmental Microbiology* **32**, 108–113.

TODD, R. L., MEYER, R. D. & WAID, J. B. 1976 Nitrogen fixation in a deciduous forest in the South Eastern United States. In *Environmental Role of Nitrogen Fixing Blue-green Algae and Asymbiotic Bacteria* ed. Granhall, U. pp.172–177. Bulletin No. 26 from the Ecological Research Committee. Stockholm: Swedish Natural Science Research Council (NFR).

VAN BERKUM, P. B. W. 1978 Nitrogenase activity associated with tropical grass roots and some effects of combined nitrogen on *Spirillum lipoferum* Beijerinck. PhD Thesis, University of London.

VLASSAK, K., PAUL, E. A. & HARRIS, R. E. 1973 Assessment of biological nitrogen fixation in grassland and associated sites. *Plant and Soil* **38**, 637–49.

VON BÜLOW, J. F. W. & DOBEREINER, J. 1975 Potential for nitrogen fixation in maize genotypes in Brazil. *Proceedings of the National Academy of Sciences, U.S.A.* **72**, 2389–2393.

VON SAUERBECK, D., JOHNEN, B. & SIX, R. 1976 Atmung, Abbau und Ausscheidungen von Weizenwurzeln im Laufe ihrer Entwicklung. *Landwirtschaftliche Forschung Sonderheft* **32**, 49–58.

WINOGRADSKY, S. 1895 Récherches sur l'assimilation de l'azote libre de l'atmosphère par les microbes. *Archives des Sciences Biologiques, St. Petersburg (Arkhiv biologicheskikh Nauk)* **3**, 297–352.

WITTY, J. F. 1979 Overestimate of N_2-fixation in the rhizosphere by the acetylene reduction method. In *The Soil-Root Interface* ed. Harley, J. L. & Scott Russell, R. pp.137–144. London & New York: Academic Press.

WITTY, J. F., DAY, J. M. & DART, P. J. 1976 The nitrogen economy of the Broadbalk Experiments. II. Biological nitrogen fixation. *Rothamsted Report for 1975* Part 2, pp.111–118.

Symbiotic Nitrogen Fixation in Plants

J. E. BERINGER

Soil Microbiology Department, Rothamsted Experimental Station, Harpenden, Hertfordshire, UK

N. BREWIN AND A. W. B. JOHNSTON

John Innes Institute, Colney Lane, Norwich, UK

Contents

1. Introduction

THE BIOLOGICAL CONVERSION of atmospheric nitrogen into ammonia is known to be accomplished only by some prokaryotic micro-organisms. The enzyme system used (nitrogenase) is apparently very similar in all species (Hausinger & Howard 1980; Ruvkun & Ausubel 1980) and has the requirement for large amounts of energy and an anaerobic environment. It is probably this large energy requirement which limits the number and significance of free-living nitrogen-fixing prokaryotes (see Ch. 2, this volume). From an agricultural point of view the most important nitrogen fixing prokaryotes are those which form symbioses with plants. The aim of this brief review is to comment on the range of these symbioses and in particular to stress some of the requirements which must be met for there to be a successful interaction between plant and micro-organism. Good reviews of various aspects of symbiotic nitrogen fixation appear in books edited by Quispel (1974), Nutman (1976), Hollaender (1977), Newton *et al.* (1977), Dobereiner *et al.* (1978), Granhall (1978) and Newton & Orme-Johnson (1980).

2. Nitrogen Fixation

A short list of prerequisites for biological nitrogen fixation is given in Table 1. The most obvious problem arising from these is that of resolving the conflicting demands of a large energy requirement and the need to

TABLE 1

Prerequisites for biological nitrogen fixation in prokaryotes

1.	The presence of a functional nitrogenase enzyme complex
2.	Large amounts of energy (5–30 mol ATP per mol N_2 reduced)
3.	Exclusion of oxygen (very rapidly inactivates nitrogenase)
4.	Fe and Mo (both required for nitrogenase)
5.	Equable temperature (above 30°C is unfavourable)
6.	Exclusion of hydrogen ions (these interact with electrons at the active site of nitrogenase releasing hydrogen gas)
7.	Regulation (to avoid waste of energy and fluctuation in ammonia production)
8.	Access to N_2

See Postgate (1974).

protect the enzyme nitrogenase from irreversible damage by oxygen. Therefore, either the micro-organism must grow anaerobically, which does not make optimum use of the energy available from a given carbon source, or it must perform oxidative phosphorylation without destroying nitrogenase. A list of oxygen-protection methods evolved by prokaryotes is given in Table 2.

Interestingly the known prokaryotes involved in symbiotic nitrogen-fixing associations with plants are aerobes, which suggests that the adaptation which must occur for the symbiosis to proceed provides a limited amount of oxygen to the micro-organisms. Specific examples of adaptations concerned with oxygen protection will be discussed later.

Hydrogen evolution by nitrogenase is a problem because it results in a direct loss of energy due to reducing power being diverted from nitrogen fixation. All nitrogenases evolve hydrogen and it appears to be an unavoidable consequence of the structure of the active site (Hageman & Burris 1978). However, different physiological environments can affect the proportion of H_2 evolved to N_2 fixed (Orme-Johnson 1977). Bacteria are able to recover some of the energy lost by hydrogen production if they possess a suitable hydrogenase (Smith *et al.* 1976). While many symbiotic associations have been shown to metabolize hydrogen (Evans *et al.* 1979) a surprising number of strains of *Rhizobium* appear not to have the necessary uptake hydrogenase(s) (Evans *et al.* 1979; Ruiz-Argueso *et al.* 1978, 1979). How important these hydrogenases are in increasing the efficiency of nitrogen fixation remains to be determined. It is an active area of research

TABLE 2
Methods evolved by prokaryotes for avoiding damage to nitrogenase by oxygen

1. Avoiding oxygen (only synthesizing nitrogenase and fixing nitrogen anaerobically)
2. Compartmentalization (e.g. some cyanobacteria only fix nitrogen in special cells, heterocysts)
3. Growing micro-aerobically
4. Having very high respiratory rates to channel O_2 from nitrogenase
5. Conformational changes in enzyme structure
6. Associating the enzyme with other macromolecules
7. Having oxygen transport systems which maintain a low pO_2

See Yates (1977).

interest at present because of the potential significance in reducing the energy requirements for symbiotic nitrogen fixation and hence increasing potential crop yields.

Symbiotic nitrogen fixation implies that the plant is providing the micro-organism with energy in return for fixed nitrogen. However the high energetic cost of biological nitrogen fixation sets a limit to the amount of fixation that a plant can support and, as a result, the population of symbiotic micro-organisms with which it can associate. The host partially controls the capacity for the micro-organisms to fix nitrogen because it determines the amount of energy available. However it appears that all the symbioses have a number of further controls on the potential for the micro-organisms to utilize this energy. Host plants seem to control very closely the number of N_2-fixing micro-organisms with which they interact, for example by enclosing them within nodules or other 'pockets' and limiting the number of such sites which are available. A striking adaptation which has been observed is the loss of ammonia-assimilating ability by rhizobia. Thus, when they start to fix nitrogen, the level of ammonia-assimilating enzymes decreases and they excrete the greater part of the nitrogen fixed as ammonia (O'Gara & Shanmugam 1976). In nodules, therefore, the ammonia-assimilating enzymes are primarily those of the plant (Brown & Dilworth 1975; Robertson *et al.* 1975). A similar situation has also been reported for the *Anabaena–Azolla* symbiosis (Peters 1977).

3. The Range of Symbiotic Nitrogen-fixing Associations

A list of these associations is given in Table 3. The rhizosphere associations discussed by Brown (Ch. 2, this volume) have not been included because, as yet, they have not been shown to be true symbioses. It is obvious from

TABLE 3

Symbiotic nitrogen fixing associations between micro-organisms and plants

Plant group	Family	Genus	Microbial symbiont	Form of association
Angiosperms	Leguminoseae	Many genera	*Rhizobium* spp.	Root nodules
	Ulmaceae	*Parasponia* (*Trema*)		
	Casuarinaceae	*Casuarina*	Actinomycetes (*Frankia*)	
	Myricaceae	*Myrica*, *Comptonia*		
	Betulaceae	*Alnus*		
	Rosaceae	*Purshia*		
	Elaeagnaceae	*Elaeagnus*, *Hippophaë*		
	Halorhagidaceae	*Gunnera*	Cyanobacteria	Various modified plant structures containing the cyanobacteria
Gymnosperms	Cycadaceae	*Cycas*, *Macrozamia*		
Pteridophytes	Salviniaceae	*Azolla*		
Fungi		Various species	Cyanobacteria	Lichens

Table 3 that a wide range of plants and micro-organisms is involved. The main link between the symbioses, other than those associated with nitrogen fixation, is that all the associations involve a complex interaction with a significant structural alteration of the host plant.

In general the symbioses involving the cyanobacteria (the blue-green algae) appear to be the most simple, though even that with fungi (forming lichens) must involve a series of complex interactions. The symbiosis between *Anabaena azollae* and the water fern *Azolla* has been studied in some detail recently because of its importance as a source of nitrogen in rice paddies in the Far East (Talley *et al.* 1977). The *Anabaena* infects the fronds of the fern at an early stage of the plant's development and becomes enclosed in a pocket within the frond (Peters 1977). Presumably the fern feeds the bacteria within these pockets and acts as the assimilatory sink for ammonia secreted by the *Anabaena*. Within these pockets the most obvious change that has occurred to the bacteria is that the heterocyst frequency is greatly increased. Heterocysts are special cells in the filament which do not contain the photosynthetic apparatus required for oxygen evolution and are the main site of nitrogenase activity. The increase in heterocyst frequency from a few per cent, when fixing nitrogen in pure culture, to 30% or more increases the nitrogen fixing potential of the limited number of *Anabaena* cells present, but coincidentally it also decreases their photosynthetic capacity. This loss of photosynthetic capacity by the *Anabaena* seems to be at variance with the observation that the symbioses are probably limited by energy supply. It might be expected that those symbioses with the cyanobacteria would be the most efficient because the microsymbiont has the potential to provide some of the required photosynthate. However, for *Azolla* it has been calculated that 80% or more of the photosynthate used for nitrogen fixation is provided by the host.

For many years it was thought that the actinomycetes which form nitrogen-fixing nodules on angiosperms were obligate symbionts since all attempts to culture them were in vain. However, recently a symbiotic actinomycete has been grown without the host and pure cultures of it have been used to induce infections in new host plants (Callahan *et al.* 1978; Burggraaf *et al.* 1981). Although these symbioses are not as important agriculturally as that between *Rhizobium* and legumes they are potentially very important to our understanding of the symbiosis because they show that micro-organisms have the ability to form nodules and fix nitrogen on the roots of plants from a number of different families (Table 3).

The best-known nitrogen fixing symbiosis is that between *Rhizobium* and leguminous plants. This is because of its agricultural significance and the fact that the rhizobia grow well and can be studied in the absence of the legume host. A brief description of this symbiosis will help to highlight some of the complex interactions which can occur between two organisms and the possible benefits that they may derive from a nitrogen fixing symbiosis.

The first stage of the interaction occurs soon after the germination of the legume seed in soil containing a *Rhizobium* species capable of nodulating it. For infection to occur the bacteria must penetrate the root and it is possibly at this level of interaction that recognition and host-*Rhizobium* specificity is mainly determined. For legumes of temperate regions, such as clover (*Trifolium* spp.), bean (*Phaseolus* spp.) and lucerne (*Medicago sativa*), there is a very tight control of the *Rhizobium* that can form nodules and the speciation of rhizobia has been based on this (Vincent *et al.* 1979). However, a significant number of 'cross-infections' have been reported and there is apparently much less specificity shown by tropical legumes and their rhizobia.

Infection of the root occurs at a limited number of sites and the plant controls the number of nodules that are formed (see Dart, 1975). This type of regulation of the extent of infection that is tolerated is observed with all legume-*Rhizobium* symbioses. It provides an obvious control of the amount of energy the plant must expend on feeding the micro-organisms and a crude control on the total amount of nitrogen which can be fixed. It is not clear how the plant recognizes the invading symbiont as benign and responds in such a controlled manner. The presence of surface antigens on *Rhizobium trifolii* which are also present on the root of the appropriate *Trifolium* host (Dazzo & Hubbell 1975) suggest that some form of deception based on common recognition antigens may be involved. However, this does not explain how the plant determines that sufficient successful infections have occurred and then prevents further infection by the same micro-organism until more nodules are required.

One of the most dramatic events during the development of a nitrogen fixing legume nodule is the production of haemoglobin which becomes a major nodule protein. It is fairly certain now that the haem is synthesized by the rhizobia (Cutting & Schulman 1972; Nadler & Avissar 1977) and the globin is produced by the plant (Sedloi-Lumbroso *et al.* 1978). In legume nodules nitrogen fixation appears to be dependent on haemoglobin. Its role is now thought to be mainly in promoting a flux of oxygen within the nodules sufficient to sustain oxidative phosphorylation by the bacteroids in the micro-aerobic conditions which favour the activity of nitrogenase. Thus the method evolved by legumes to overcome the energy and oxygen damage problem posed at the beginning of this paper involves both partners.

Haemoglobin has not been convincingly detected in nodules formed by the actinomycete *Frankia* in a non-legume, or in the nodules formed by *Rhizobium* on the non-legume *Parasponia* (formerly *Trema*) (Coventry *et al.* 1976). How the actinomycete nodules handle the oxygen problem has yet to be clarified. The rhizobia which nodulate *Parasponia* can form normal nitrogen fixing nodules on legume hosts and are also able to fix nitrogen in the *Parasponia* nodules. It appears that they are not released from the infection threads in which they invade the root cortex (Dart 1975). Thus a sufficiently low pO_2 for nitrogenase activity probably exists. It should be possible to compare the efficiency of nitrogen fixation in nodules formed on *Parasponia* and cowpeas by the same *Rhizobium* strain to determine the proposed role of haemoglobin in increasing the efficiency of nitrogen fixation in nodules. This does not appear to have been done.

The nodulation of *Parasponia* by *Rhizobium* shows that there is nothing unique about the genetic structure of legumes that enables them to develop a nitrogen fixing symbiosis with *Rhizobium* and also that the actinomycetes are not unique in being able to nodulate and fix nitrogen in a range of plant families. It may well be that many dicotyledonous plants have the necessary genes to allow controlled microbial infections and the development of nodules in which these infections can be contained. A greater understanding of these symbioses should help us to know whether the range of nodulated nitrogen fixing plant species can be extended in the future.

4. References

BROWN, C. M. & DILWORTH, M. J. 1975 Ammonia assimilation by *Rhizobium* cultures and bacteroids. *Journal of General Microbiology* **86**, 39–48.

BURGGRAAF, A. J. P., QUISPEL, A., TAK, T. & VALSTAR, J. 1981 Methods of isolation and cultivation of *Frankia* species from actinorhizas. *Plant and Soil* in press.

CALLAHAN, D., DEL TREDICI, P. & TORREY, J. G. 1978 Isolation and cultivation *in vitro* of the actinomycete causing root nodulation in *Comptonia*. *Science, N.Y.* **199**, 899–902.

COVENTRY, D. R., TRINICK, M. J. & APPLEBY, C. A. 1976 Search for a leghaemoglobin-like compound in root nodules of *Trema cannabina* Lour. *Biochimica et Biophysica Acta* **420**, 105–111.

CUTTING, J. A. & SCHULMAN, H. M. 1972 The control of heme synthesis in soybean root nodules. *Biochimica et Biophysica Acta* **261**, 321–327.

DART, P. J. 1975 Legume root nodule initiation and development. In *The Development and Function of Roots* ed. Torrey, J. G. & Clarkson, D. T. pp.467–506. London & New York: Academic Press.

DAZZO, F. B. & HUBBELL, D. H. 1975 Cross-reactive antigens and lectins as determinants of symbiotic specificity in the *Rhizobium*-clover association. *Applied Microbiology* **30**, 1017–1033.

DOBEREINER, J., BURRIS, R. H. & HOLLAENDER, A. 1978 *Limitations and Potentials for Biological Nitrogen Fixation in the Tropics* ed. Dobereiner, J., Burris, R. H. & Hollaender, A. New York: Plenum Press.

EVANS, H. J., EMERICH, D. W., MAIER, R. J., HANUS, F. J. & RUSSELL, S. A. 1979 Hydrogen cycling within the nodules of legumes and non-legumes and its role in nitrogen fixation. In *Symbiotic Nitrogen Fixation in the Management of Temperate Forests* ed. Gordon, J. C., Wheeler, C. T. & Perry, D. A. pp.196–206. Oregon State University Press.

GRANHALL, U. 1978 *Environmental Role of Nitrogen-Fixing Blue-green Algal and Asymbiotic Bacteria* ed. Granhall, U. Ecological Bulletin 26 (Stockholm).

HAGEMAN, R. V. & BURRIS, R. H. 1978 Nitrogenase and nitrogen reductase associate and dissociate with each catalytic cycle. *Proceedings of the National Academy of Sciences, U.S.A.* **75**, 2699–2702.

HAUSINGER, R. P. & HOWARD, J. B. 1980 Comparison of the iron protein from the nitrogen fixation complexes of *Azotobacter vinelandii*, *Clostridium pasteurianum* and *Klebsiella pneumoniae*. *Proceedings of the National Academy of Sciences, U.S.A.* **77**, 3826–3830.

HOLLAENDER, A. 1977 *Genetic Engineering for Nitrogen Fixation* ed. Hollaender, A. New York: Plenum Press.

NADLER, K. D. & AVISSAR, Y. J. 1977 Heme synthesis in soybean root nodules. I. On the role of bacteroid δ-aminolevulinic acid synthase and δ-aminolevulinic acid dehydrase in the synthesis of the heme of leghaemoglobin. *Plant Physiology* **60**, 433–436.

NEWTON, W. E. & ORME-JOHNSON, W. H. 1980 *Nitrogen Fixation* Vol. 2 *Symbiotic Associations and Cyanobacteria* ed. Newton, W. E. & Orme-Johnson, W. H. Baltimore: University Park Press.

NEWTON, W. E., POSTGATE, J. R. & RODRIGUEZ-BARRUECO, C. 1977 *Recent Developments in Nitrogen Fixation* ed. Newton, W. E., Postgate, J. R. & Rodriguez-Barrueco, C. London & New York: Academic Press.

NUTMAN, P. S. 1976 *Symbiotic Nitrogen Fixation* ed. Nutman, P. S. Cambridge: Cambridge University Press.

O'GARA, F. & SHANMUGAM, K. T. 1976 Regulation of nitrogen fixation by rhizobia. Export of fixed N_2 as NH_4^+. *Biochimica et Biophysica Acta* **437**, 313–321.

ORME-JOHNSON, W. H. 1977 The biochemistry of nitrogenase. In *Genetic Engineering for Nitrogen Fixation* ed. Hollaender, A. New York: Plenum Press.

PETERS, G. A. 1977 The *Azolla-Anabaena azolae* symbiosis. In *Genetic Engineering for Nitrogen Fixation* ed. Hollaender, A. New York: Plenum Press.

POSTGATE, J. R. 1974 Prerequisites for biological nitrogen fixation in free-living heterotrophic bacteria. In *The Biology of Nitrogen Fixation* ed. Quispel, A. Amsterdam: North-Holland.

QUISPEL, A. ed. 1974 *The Biology of Nitrogen Fixation*. Amsterdam: North-Holland.

ROBERTSON, J. G., WARBURTON, M. P. & FARNDEN, K. J. F. 1975 Induction of glutamate synthase during nodule developent in lupin. *Federation of European Biochemical Societies Letters* **55**, 33–37.

RUIZ-ARGUESO, T., HANUS, F. J. & EVANS, M. J. 1978 Hydrogen production and uptake by pea nodules as affected by strains of *Rhizobium leguminosarum*. *Archives of Microbiology* **116**, 113–118.

RUIZ-ARGUESO, T., MAIER, R. J. & EVANS, H. J. 1979 Hydrogen evolution from alfalfa and clover nodules and hydrogen uptake by free-living *Rhizobium meliloti*. *Applied and Environmental Microbiology* **37**, 582–587.

RUVKUN, G. B. & AUSUBEL, F. M. 1980 Interspecies homology of nitrogenase genes. *Proceedings of the National Academy of Sciences, U.S.A.* **77**, 191–195.

SEDLOI-LUMBROSO, R., KLEIMAN, L. & SCHULMAN, H. M. 1978 Biochemical evidence that leghaemoglobin genes are present in the soybean but not the *Rhizobium* genome. *Nature, London* **273**, 558–560.

SMITH, L. A., HILL, S. & YATES, M. G. 1976 Inhibition by acetylene of conventional hydrogenase in nitrogen-fixing bacteria. *Nature, London* **262**, 209.

TALLEY, S. N., TALLEY, B. J. & RAINS, D. W. 1977 Nitrogen fixation by *Azolla* in rice fields. In *Genetic Engineering for Nitrogen Fixation* ed. Hollaender, A. New York: Plenum Press.

VINCENT, J. M., NUTMAN, P. S. & SKINNER, F. A. 1979 The identification and classification of *Rhizobium*. In *Identification Methods for Microbiologists* 2nd edn, ed. Skinner, F. A. & Lovelock, D. W. London & New York: Academic Press.

YATES, M. G. 1977 Physiological aspects of nitrogen fixation. In *Recent Developments in Nitrogen Fixation* ed. Newton, W. E., Postgate, J. R. & Rodriguez-Barrueco, C. New York: Academic Press.

Entry and Establishment of Pathogenic Bacteria in Plant Tissues

EVE BILLING

East Malling Research Station, Maidstone, Kent, UK

Contents

1. Introduction

MUCH CURRENT RESEARCH is focussed on questions of host–pathogen interactions during the early stages of infection in bacterial plant diseases (Goodman 1976; Kelman 1979; Chatterjee & Starr 1980). Space does not allow a comprehensive critical review of such work and only an outline of current concepts is possible with selected references by way of illustration. While this will do less than justice to the workers in this field, it may provide the reader with a base on which to build when studying accounts of recent research.

Plant pathogenic bacteria described in most detail here include those which produce necrotic symptoms in blossoms, leaves and stems (blights, leaf spots and cankers) and wilting. Pectolytic bacteria which produce soft rots of parenchyma tissue and those producing tissue proliferation are mentioned only briefly as they are the subject of other contributions in this volume (Chs 7 & 9).

Plant pathogens discussed are confined to three Gram negative genera: *Pseudomonas*, *Xanthomonas* and *Erwinia* and the Gram positive coryneform bacteria (Buchanan & Gibbons, 1974). The taxonomy of plant associated bacteria is described by Billing (1976) and new nomenclatural recommendations are made by Dye *et al.* (1980).

2. Entry into Host Tissue

Pathogenic bacteria are not equipped to penetrate intact plant surfaces and entry is via natural openings and wounds. Young tissue is much more susceptible than mature tissue and water-soaking greatly increases chances of infection as it allows free passage of bacteria from outer to inner surfaces. Chemotaxis has been observed so functional flagella are important to the pathogen at this stage (Panopoulos & Schroth 1974; Feng & Kuo 1975; Goodman 1976; Schmit 1978).

In blossoms, initial multiplication may occur on the stigma and in the receptacle with subsequent invasion via nectarial or other tissue (Hildebrand 1937; Thomson 1978). Entry may also occur via stomata in the calyx (Miller 1929; Panagopoulos & Crosse 1964).

With shoots, entry is usually via unfolding and expanding leaves near the tip. In leaf spot and other necrotic diseases, bacteria commonly enter via stomata (Panopoulos & Schroth 1974) and multiply initially in the substomatal chamber whilst in some vascular diseases, entry is via hydathodes with initial multiplication in subtending cortical tissue prior to invasion of xylem vessels (Staub & Williams 1972; Tabei 1973).

In stems, including potato tubers (Lund 1979), entry may be via young lenticels and, in woody plants, via leaf scars which can be particularly vulnerable if leaves are removed prematurely (Crosse 1956; Feliciano & Daines 1970). A weak point in roots is where secondary roots emerge and *Pseudomonas solanacearum* may multiply there prior to invasion of cortical tissue from where subsequent entry to xylem vessels is aided by tylose formation (Schmit 1978; Wallis & Truter 1978).

Wounds are of major importance in infection and may permit direct entry to xylem vessels. They may be induced by wind, hail, insects (Leach 1940), nematodes Pitcher 1963) or birds. Pruning cuts and damage during cultivation or harvesting can also be important. The plant is most vulnerable when damage and water congestion coincide such as occurs during storms. Wind driven rain can simultaneously spread inoculum and aid water congestion whilst strong winds or hail induce damage; torn leaves with their vascular system exposed are particularly vulnerable to infection (Crosse *et al.* 1972; Daft & Leben 1972).

3. Survival and Growth of Bacteria in the Intercellular Environment

A. Growth requirements of plant pathogens

Most necrosis and wilt bacteria are strict aerobes. Of the facultative anaerobes, *Erwinia amylovora* grows poorly anaerobically but the soft-rotting *E. carotovora* grows well. Few pathogens are nutritionally exacting though some require growth factors or amino acids. Most grow well in the range pH 6·5–7·5, some show reduced growth below pH 6·0 and few will grow below pH 4·5. Like other bacteria their water requirement for optimal growth is high (above 99% relative humidity).

B. The plant tissue environment

The most common sites of multiplication are either the intercellular spaces of young parenchyma tissue or xylem vessels. Most bacteria which primarily invade parenchyma also multiply at times in xylem vessels, though they may have difficulty in escaping, and vascular pathogens may have to traverse parenchyma tissue to reach xylem vessels.

(i) *Nutrients*

Because of sampling difficulties, it is difficult to judge pH and nutrient composition of fluids at these sites. Young plant tissue will be importing nutrients and bacteria there should be well supplied, but nutrient levels in xylem vessels may be low compared with those in intercellular spaces.

(ii) *Inhibitors*

Preformed inhibitors may be present in bound or free form. Some are present as glycosides and may be liberated by β-glycosidases of plant or bacterial origin, others can arise from oxidation of phenolic substances released as a result of damage or infection. To be effective, they must be present in free form and in high enough concentration at the site of infection; as with nutrients, actual levels are difficult to judge and evidence of the role of preformed inhibitors in resistance is sparse and often inconclusive (Schönbeck & Schlösser 1976). Amino acids can be inhibitory under some circumstances and it has been suggested that their nature and balance may be contributory factors in host specificity (Sands & Zucker 1976).

The balance between inhibitors and stimulatory nutrients at points of entry and initial multiplication may also be a delicate one and part of the basis for host specificity may lie here. Extracts from plants sometimes appear more inhibitory for incompatible than for compatible pathogens (Kelman & Sequeira 1972) and chemotactic studies suggest that exudates from resistant plants may be less attractive than those from susceptible ones (Feng & Kuo 1975) though this is not always the case (Goodman 1976).

(iii) *Water-soaking*

Intercellular spaces of plants normally contain air but under certain conditions the spaces fill with water and such water congestion greatly favours infection (Johnson 1947). Factors favouring congestion are: innate proneness of the plant; a well-aerated, warm moist soil with high N and low K levels; soil temperatures higher than air temperatures; plant surface wetness and high humidity. These conditions favour water uptake and transport within the plant and stomatal opening (Johnson 1947; Matthee & Daines 1968). In uncongested tissue the pathogen must rely at first on water in plant cell walls but the art of a successful pathogen is to induce an increase in membrane permeability of the host cell and to retain the water thus leaked. Characteristically as infection progresses there is persistent water-soaking at sites of bacterial multiplication, prior to cell collapse and necrosis.

C. Experimental observations

Quantitative studies of several diseases indicate that bacterial cells act independently and that, under favourable conditions, a single cell may initiate infection (Crosse *et al.* 1972; Ercolani 1973; Essenberg *et al.* 1979*a*).

Plant–bacterium interactions vary. Those commonly observed are summarized in Tables 1 and 2, but most discussion will centre round interactions in Table 1. Many studies of interactions involve infiltration of bacteria into intercellular spaces of leaves and the findings of Ercolani & Crosse (1966) and Young (1974*a*, *b*) with fluorescent *Pseudomonas* species in cherry and bean leaves are sufficiently representative of those of other workers to serve as a model though the precise outcome can vary with different host–bacterium systems or different experimental conditions.

The picture that emerges is that the compatible pathogen multiplies freely and induces visible water congestion of tissue when it reaches the stationary phase in about four days. The incompatible pathogen, if it grows, does so at a slower rate and growth stops abruptly after about two days so high populations are not achieved though numbers may not decline rapidly. Non-pathogens fail to multiply and decline slowly in numbers.

If, following infiltration, leaves are kept water-soaked, incompatible

TABLE 1
Plant-bacterium interactions—pathogens causing leaf spots, blights and cankers (necrosis) and wilts

Host plant	Virulence of pathogen*	System	Usual outcome
Susceptible	+	Compatible	Progressive disease
Susceptible	-	Incompatible	Hypersensitive (HR) or other response; bacteria stop multiplying
Resistant cultivar†	+/-		
Non-host	+/-		

*Capacity to produce disease in a susceptible cultivar of the homologous host; -, avirulent strain of a virulent pathogen (some avirulent strains act like non-pathogens, Table 2).
†Or resistant portion of a susceptible host.
Data from Kelman & Sequeira (1972).

TABLE 2
Plant-bacterium interactions—pathogens causing soft rot and galls and non-pathogens

Host plant	Characteristics of bacteria	System	Typical outcome
Susceptible	Pectolytic	Compatible	Progressive soft rot disease
Many plants	Pectolytic	Opportunist*	Soft rot
Many plants (dicotyledons)	*Agrobacterium tumefaciens* with Ti plasmid	Plasmid transfer	Crown gall
Any plant	Non-pathogens	Inhibitory	No disease

*Can exploit situations where there is damaged tissue and high relative humidity.

pathogens reach higher populations and non-pathogens may multiply to a limited extent (Young 1974*b*).

Thus it appears that growth of a non-pathogen can be strongly inhibited at an early stage but that of an incompatible pathogen sometimes only slightly so initially; the failure of the latter to reach high populations seems to be largely due to a delayed defence reaction. The compatible pathogen shows little or no inhibition initially, fails to induce other defence reactions and can draw on the host cell's resources for water and nutrients. An examination of mechanisms which may be involved follows.

4. The Interacting Surfaces in Intercellular Spaces

The bacterial pathogen is small in comparison with the plant cell (Fig. 1) but is none the less a potent force. There appears to be no direct contact

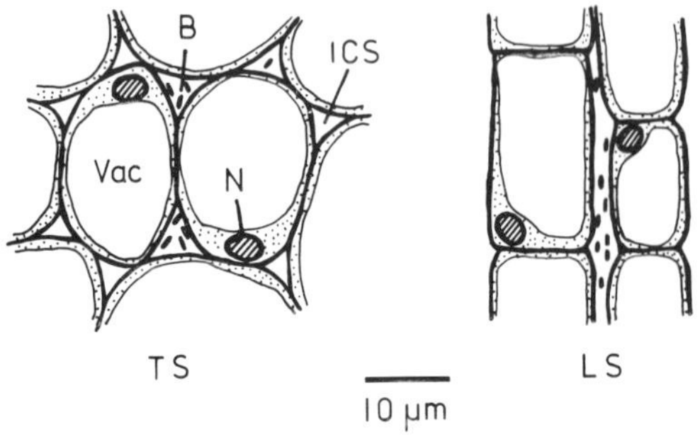

Fig. 1. Diagram of parenchyma tissue sections showing bacteria (B) in intercellular spaces (ICS). N, nucleus; Vac, vacuole. Scale approximate.

between plant and bacterial cell membranes because between them lie the plant cell wall and the bacterial cell envelope plus any extracellular polysaccharide. These coverings are far from impervious however and it is evident that one cell can rapidly have a profound effect upon the other.

A. The primary plant cell wall and membranes

The primary cell wall is a dynamic structure which changes in composition and properties as growth proceeds. It is made up of cellulose fibrils and a matrix containing polysaccharides, including pectic substances and hemicelluloses, and proteins with a glycoprotein rich in hydroxyproline predominant; some proteins have properties of lectins (see below). The water content varies, depending on cell age, the composition of the matrix and on environmental conditions. The middle lamella which cements cell walls together is composed primarily of pectic substances (Bateman & Basham 1976).

The cytoplasmic membrane (plasmalemma) is believed to resemble animal and bacterial cell membranes and to be composed largely of lipids, especially phospholipids and also glycolipids and structural and enzymic proteins including glycoproteins (Bateman & Basham 1976; Callow 1977). The plant cytoplasm contains a variety of membrane-bound organelles including the nucleus, mitochondria, Golgi apparatus and chloroplasts; another membrane, the tonoplast, encloses a large vacuole.

Lectins are carbohydrate binding proteins and occur in animal and microbial cells as well as in plant cells. Some plant lectins are firmly bound to cell walls or membranes but others can be readily extracted. Free lectins often have multiple binding sites and so may agglutinate bacteria or precipitate polysaccharides, glycoproteins or glycolipids. They can be highly specific but different types of cell (e.g. bacteria, fungi and red blood cells) may share similar receptors. In plant disease, lectins are thought to have a role in recognition and binding but they may not be sufficiently variable and specific to account for all types of specificity observed in plant disease. The subject is reviewed in detail by Callow (1977) and Sequeira (1978).

B. Bacterial surface structures and membranes

Current concepts of the structure of the cell envelope of the Gram negative bacterium *Escherichia coli* are shown in Fig. 2. How far it is representative

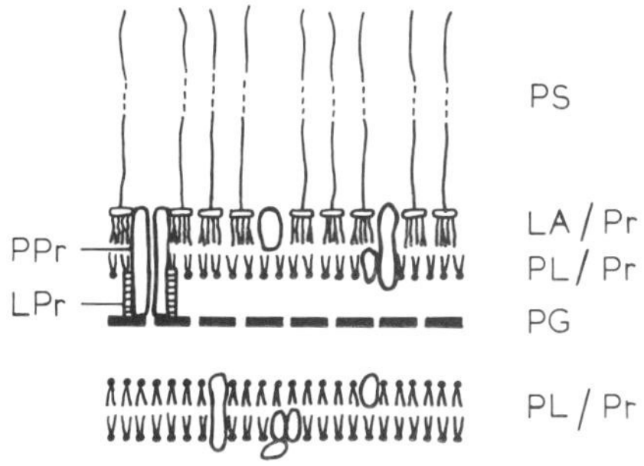

Fig. 2. Structural components of the Gram negative cell envelope based on tentative proposals of Nkaido & Nakae and others in Inouye (1979) for *Escherichia coli*. PS and LA represent the oligosaccharide chains and Lipid A portions of the lipopolysaccharide respectively; Pr, proteins; PPr, porin protein; LPr, lipoprotein; PL, phospholipid; PG, peptidoglycan.

for plant pathogens is uncertain. In Gram positive bacteria, the peptidoglycan layer is much thicker and lipopolysaccharide (LPS) is lacking, but other surface polymers are present which, like LPS side chains, are flexible, hydrophilic and negatively charged molecules which extend out from cell walls and may be specific binding sites for phages and antibodies (see Sutherland 1977). LPS is made up of Lipid A (the biologically active region of the endotoxin important in animal disease), core polysaccharide, and oligosaccharide side chains which can be changed or lost without affecting viability of the cell. They are responsible for the wide variation in antigenic specificity observed in Gram negative bacteria.

Walls may be covered by a capsule or loose polysaccharide slime (EPS). Some bacteria have projecting protein threads with lectin-like properties (pili or fimbriae). Probably all the phytopathogens have flagella and are motile.

When cells are plasmolysed, portions of cytoplasmic membrane adhere to the cell wall. These zones of adhesion are areas of uptake of nutrients and export of cell wall material and sites of injection of phage nucleic acid and of bacteriocin adsorption; chromosomal DNA is seen attached to such sites and they might well be important in host pathogen interactions. A full account of current concepts of bacterial envelope and membrane structure is given by Inouye (1979). Key features of this outline description are summarized in Fig. 3.

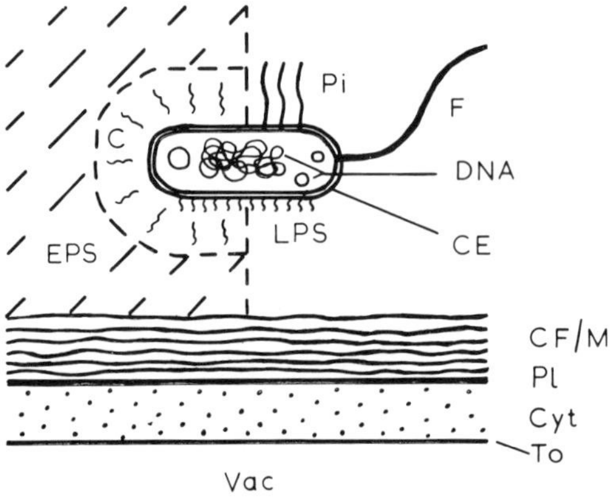

Fig. 3. Diagram of components of a bacterial cell shown in association with a plant cell (organelles omitted). EPS, extracellular polysaccharide; C, capsule; LPS, lipopolysaccharide side-chains (or other polymers in Gram positive bacteria); Pi, pili; F, flagellum; DNA, chromosomal or plasmid DNA; CE, cell envelope; CF/M, cellulose fibrils and matrix of plant cell wall; P1, plasmalemma; Cyt, cytoplasm; To, tonoplast; Vac, vacuole.

5. Host–Bacterium Interactions

A. Agglutination

Bacterial cells may be agglutinated by macromolecules with multiple binding sites for specific cell components, but non-specific clumping may also occur, especially with cells that are deficient in surface polysaccharides and hence more hydrophobic.

Avirulent cells of *Ps. solanacearum* which lack extracellular polysaccharide (EPS) are agglutinated by lectins from potato and tobacco (Sequeira & Graham 1977). Fluorescent saprophytic and phytopathogenic pseudomonads are agglutinated by lectins from various sources and also by pectins and galacturonic acid (Anderson & Jasalavich 1979). Some strains of *E. amylovora* lacking EPS agglutinate in Indian ink (Billing 1960) and in the presence of EPS from virulent strains (Hsu & Goodman 1978).

Preliminary evidence suggests the possibility that avirulent or incompatible pathogens may be agglutinated within plant tissue and that prior exposure to such bacteria may result in an increase in agglutinins whose nature is uncertain (Goodman *et al.* 1976*a*).

B. Binding

Experiments where tobacco leaves were infiltrated with bacteria suggested that incompatible and avirulent strains of compatible bacteria were rapidly bound to cell walls and engulfed by material largely of cell wall origin; non-pathogens were either loosely or firmly bound but compatible pathogens remained free (Goodman *et al.* 1976*b*; Sequeira *et al.* 1977; Sequeira 1978).

Workers using bean leaves have not always obtained comparable results. For example, Hildebrand *et al.* (1980) observed no difference between a compatible pathogen and incompatible pathogens and non-pathogens 3 h after infiltration. They suggested that some of the engulfing structures observed in tobacco and other hosts might be related to materials dissolved from cell walls during infiltration which formed films as the intercellular fluid in the leaves dried; compatible pathogens would appear free at later stages because of their ability to multiply and escape from the films.

Tobacco callus culture work presented another picture (Huang & van Dyke 1978). Here an incompatible pathogen multiplied but formed aggregates and became surrounded by a thick fibrillar network in 24–36 h whereas the compatible pathogen remained free; a non-pathogen also remained free but multiplied little. In rice leaves where incompatible bacteria multipled in xylem vessels, fibrillar material (apparently of plant origin) was seen in vessels in 3 days in the incompatible system but not for 20 days in the compatible one (Horino 1976).

Thus it seems that fibrillar material of plant origin may accumulate in incompatible interactions but whether its production is initiated before induction of a hypersensitivity reaction (HR, see below) is unclear. Since bacteria will be drawn towards plant cell walls as intercellular spaces dry out, it seems unnecessary to postulate a specific binding process, though this may occur. The failure of incompatible bacteria to progress following initial multiplication could be due in part to their immobilization by fibrillar material and agglutination, but consequences of HR induction must also be considered.

C. Hypersensitivity reaction

The hypersensitivity reaction (HR) in plants is distinct from that described for animals where sensitization, due to prior exposure to an antigen, is involved. In plants it refers to tissue damage and other effects resulting from exposure to an incompatible pathogen (Kelman & Sequeira 1972; Sequeira 1976, 1978). The main events are: rapid electrolyte leakage, loss of cell turgor, desiccation and necrosis; inhibitory substances also accumulate. The process may be irreversible 3 h after inoculation and cell collapse may be complete in 12 h. An HR test is usually made by infiltrating bacteria into intercellular spaces of tobacco leaves; results with other host plants may differ in detail and in timing but the overall reaction is comparable. Apart from occasional delayed reactions, non-pathogens produce little damage. The compatible pathogen produces delayed electrolyte leakage followed by water-soaking and progressive infection.

Current evidence suggests that the HR inducing agent is only produced by live, metabolically active bacteria with intact cell walls, and protein synthesis by the bacteria appears necessary (Sequeira 1976; Sasser 1978). Sustained water-soaking of leaves or infiltration with a 0·5% agar gel delays or prevents HR as well as allowing increased multiplication of incompatible pathogens and non-pathogens (Young 1974*b*; Stall & Cook 1979). This is suggested as evidence that contact of bacteria with cell walls is necessary for HR induction. Hildebrand *et al.* (1980) point out that water-soaking need not prevent an active binding process.

Ercolani (1973) suggested that bacteria acted co-operatively to induce HR but other evidence indicates that they act independently (Sequeira 1976; Essenberg *et al.* 1979*a*, *b*). Rapid death of incompatible bacteria following HR seems to be the exception rather than the rule (Ercolani & Crosse 1966; Essenberg *et al.* 1979*b*); their failure to progress in plant tissue may be attributable to failure to exploit host cell water and nutrient resources in a controlled manner as well as to inhibition or immobilization.

Antibiotic substances known as phytoalexins accumulate in necrotic regions in response to continuous stimulus by physical, chemical or biological agents and to infection, but their relationship to resistance in bacterial diseases in unclear (Kuć 1976; Bruegger & Keen 1979).

6. Roles of Bacterial Cell Components

A. Lipopolysaccharides (LPS)

Host specificity is sometimes correlated with the presence of specific antigens or sensitivity to certain phages but when such attributes are sought, success is limited (Trigalet *et al.* 1978; Billing & Garrett 1980). Some such antigens or phage receptors are LPS in nature (Lucas & Grogan 1969; Quirk *et al.* 1976), but as will be seen, the role of LPS in host specificity is not clear.

The idea that LPS may be concerned in host specificity is an attractive one because of the scope for variation in oligosaccharide side chains without effect on viability of the pathogen (Albersheim & Anderson-Prouty 1975; Sequeira 1978). Lectins are obvious binding sites in the host cell though there is evidence that *Agrobacterium tumefaciens* LPS binds to pectic substances (Lippincott & Lippincott 1977; see also Section 5.A).

Experience with *Ps. solanacearum* suggests that lectin–LPS interactions may not be highly specific. Potato lectin agglutinated avirulent strains of all three races and the same was true of virulent strains when EPS was removed by washing. The lectin bound to LPS but it was the Lipid A fraction not the oligosaccharide side chains that appeared to be the binding site (Sequeira & Graham 1977).

In tobacco, resistance to HR and infection can be induced by pre-infiltration with living or heat-killed bacteria or LPS-containing extracts. With *Ps. solanacearum*, the inducing agent, like the lectin receptor, appears to be Lipid A (Sequeira 1976, 1979) but with *Ps. morsprunorum* an association with protein seems necessary (Quirk *et al.* 1976) and the same may be true of other fluorescent *Pseudomonas* spp. (Mazzuchi *et al.* 1979). With other hosts, resistance induction has not been so readily achieved (Essenberg *et al.* 1979*a*) and other agents can induce resistance. The complexities of the subject of induced resistance are discussed by Kuć (1979) and Sequeira (1979).

With *Ps. morsprunorum* there is further evidence that for full biological activity (toxicity, host specificity and phage sensitivity) a protein–LPS complex may be necessary (Hignett & Quirk 1979).

B. Extracellular polysaccharides (EPS)

EPS production may be a universal characteristic of bacterial plant pathogens and is sometimes associated with capsulation (Billing 1960; Horino 1976; Bennett & Billing 1978). Polysaccharides produced *in vitro* appear to be similar to those produced *in vivo* (*Xanthomonas campestris*, Sutton & Williams 1970; *X. oryzae*, Angadi 1978; *E. amylovora*, Eden-Green & Knee 1974; Bennett & Billing 1980). *Erwinia amylovora* was once thought to produce a specific polysaccharide only *in vivo* (Goodman *et al.* 1974; Strobel 1977).

Loss of ability to produce EPS is associated with loss of virulence (*Ps. solanacearum*, Husain & Kelman 1958; *X. campestris*, Sutton & Williams 1970; *E. amylovora*, Bennett & Billing 1978; Ayers *et al.* 1979) but the existence of a capsulated EPS-producing strain of *E. amylovora* which is avirulent indicates that EPS production is not the sole determinant of virulence in this pathogen (Bennett & Billing 1980).

The chemical composition of EPS produced by pathogens varies as widely as with those produced by other bacteria (Strobel 1977; Sutherland 1977) but most are hydrophilic, acidic heteropolymers. EPS associated with nitrogen residues, as in *Corynebacterium* spp., may be referred to as glycopeptides (Strobel 1977). Some pathogens (e.g. fluorescent pseudomonads and *E. amylovora*) produce in addition a homopolymer levan (α-poly-D-fructan) in the presence of 5% sucrose (Lelliott *et al.* 1966; Bennett & Billing 1980); its relation, if any, to virulence is not known.

It seems likely that EPS has multiple roles in infection. Roles attributed to bacterial surface polysaccharides in natural environments include attraction of water and nutrients and protection of the cell and its enzymes from adverse environmental factors (see Sutherland 1977). In animal disease, the hydrophilic nature of EPS protects cells from phagocytosis; in plant disease it appears to interfere with binding of LPS to host lectins and to suppress the HR. It is easy to envisage other ways in which this hydrophilic surface material might be an advantage to the pathogen, including facilitating entry to and progression through intercellular spaces and xylem vessels.

There is evidence of specific interactions of possible importance in establishment of infection. EPS from *X. campestris* (xanthan) and *X. phaseoli* forms gels when mixed with plant cell wall polysaccharides with β-1,4-linked backbones (Powell 1979). Xanthan has unusual rheological properties and might be a special case but the concept of such an interaction between host and pathogen polysaccharides is an attractive one. The reaction described does not suggest a basis for host specificity but there is other evidence that EPS may interact specifically with host tissue components. El Banoby & Rudolph (1979) found that infiltration of EPS into leaves in compatible (but not incompatible) host pathogen combinations, with subsequent incubation at high humidity, resulted in persistence of water-soaking in infiltrated areas. Progressive persistent water-soaking is a feature of compatible systems, and whatever induces the condition, it seems logical that the EPS should have a role in its persistence. If for this a stable gel is necessary, compatibility of host and pathogen polysaccharides could be important.

Another indication that host specificity may in part lie with the composition of its EPS is that possession of EPS does not itself prevent an incompatible strain from inducing HR in a non-host plant, though it does protect it from agglutination by lectins (Sequeira & Graham 1977). A parallel situation has been seen in a fungal disease where glucans from a compatible race were more active in supressing HR than those from incompatible races (Doke *et al.* 1979).

(i) *Wilt induction by EPS*

EPS may have a role in the wilting process in the later stages of infection in vascular wilt diseases and this has led to the concept that the ability of an EPS to induce wilting in cut shoots is an indication of its toxicity. There is increasing evidence, however, that this ability is largely a function of the molecular weight of the polysaccharide and that susceptibility of shoots to wilting by macromolecules is largely a function of their vascular anatomy, so it can be confusing to refer to them as phytotoxins (Hodgson *et al.* 1949; Van Alfen & Allard-Turner 1979). Anatomy may be important, therefore, where results of wilt tests appear to reflect host susceptibility (e.g. Goodman *et al.* 1974); in Dutch elm disease, resistance appears to be related to a vascular anatomy which restricts water movement (Ayres 1978).

Wilt induction has been reported for EPS from: *A. tumefaciens* (Hodgson *et al.* 1949); *P. solanacearum* (Husain & Kelman 1958); three *Xanthomonas* spp. (Powell 1979; Sutton & Williams 1970; Angadi 1978), three *Corynebacterium* spp. (Strobel 1977) and *E. amylovora* (Goodman *et al.* 1974). With EPS from *E. amylovora* (amylovorin) the original estimate of a molecular weight of $1{\cdot}65 \times 10^5$ appears too low (Goodman *et al.* 1974; Strobel 1977; Bennett & Billing 1980) and it probably exceeds 1×10^6. The EPS seems to induce wilt by restriction of water movement and dextran of comparable molecular weight is equally effective (Sjulin & Beer 1978). Ultrastructural studies suggest that the EPS induces host cell plasmolysis similar to that seen at an early stage in natural infections (Huang & Goodman 1976); the authors considered this a toxic effect rather than evidence of water stress. It could be argued that EPS has a role in delaying electrolyte leakage rather than inducing membrane damage.

The glycopeptide produced by *Corynebacterium insidiosum* with a molecular weight of *ca.* 5×10^6 also induces wilt by impairment of water movement (Strobel 1977). Glycopeptides from *C. michiganense* and *C. sepedonicum*, however, have lower molecular weights (Strobel 1977) and might only accumulate at leaf margins. With the latter, membrane damage to stem cells of cut shoots was detected suggesting a direct toxic effect (Strobel & Hess 1968) but glycopeptide concentrations used seem high (10 mg/ml) especially if an effect in the early stages of infection is under consideration.

C. Other cell components

(i) *Flagella*

These appear to be necessary for entry of bacteria to initial multiplication sites but not thereafter (Panopolous & Schroth 1974; Schmit 1978; Essenberg *et al.* 1979*a*).

(ii) *Pili*

Pili have rarely been reported for plant pathogens and have no known role in plant disease. In animal disease, specific adherence may be important in location of multiplication sites and to prevent the pathogen being swept away by body fluids; even there, LPS and teichoic acid are thought to have a role also in recognition and adhesion (see Ellwood *et al.* 1979). In plant disease, adherence might be a disadvantage (*A. tumefaciens* apart) and limit rates of progression.

(iii) *Peptidoglycans and phospholipids*

Since components of peptidoglycans are unique to bacteria, it is surprising that there is no evidence that plants react to them. Evidence of interactions involving phospholipids is also lacking.

(iv) *Proteins*

There is undoubtedly much to learn about proteins in relation to plant–bacterium interactions but soluble lectins apart, they are less amenable to study than are polysaccharides.

In a hypothesis concerned with specificity in both fungal and bacterial diseases, Albersheim & Anderson-Prouty (1975) suggested that a host resistance gene codes for a protein located in the host plasma membrane, and that the relevant gene in the pathogen also codes for a protein which they suggest is a glucosyl transferase enzyme concerned with synthesis of cell surface carbohydrates. It was postulated that the host cell protein acts as a receptor for a surface carbohydrate in the incompatible system and that the interaction results in a resistance reaction; in the compatible system the receptor fails to interact. Callow (1977) discussed a similar system.

Vanderplank (1978) considered an interaction involving two chemically different types of molecule as unlikely on biochemical and thermodynamic grounds, and proposed that interactions involve co-polymerization of host and pathogen proteins. In the compatible system, this might result in new protein synthesis and inactivation of the catalytic action of the pathogen's protein, thus preventing a resistance reaction. In the incompatible system, the proteins perhaps fail to interact and the catalytic action of the pathogen's protein induces a resistance reaction. Both hypotheses are concerned with gene-for-gene systems and are tentative; neither host nor pathogen protein has been identified. Other evidence suggests additional possibilities and disease resistance appears likely to involve a variety of mechanisms (Kuć 1979; Dixon & Lamb 1980).

(v) *Common antigens*

There are instances where a bacterial pathogen has antigens in common with its host plant (De Vay & Adler 1976); it was suggested that in such cases a plant would be more tolerant of an invading organism or that plant lectins might form a link between common receptors on host and pathogen. The overall significance of antigen sharing is still an open question.

(vi) *Enzymes, toxins and growth regulators*

Physiologically active compounds produced by bacterial plant pathogens include hydrolytic enzymes (e.g. cellulases, pectinases, proteases, lipases, glycosidases and phosphatases), plant growth regulators (indole acetic acid, cytokinins and ethylene) and low molecular weight toxins (e.g. syringomycin, tabtoxin and phaseolotoxin). Unlike some animal pathogens, bacterial plant pathogens do not appear to produce extracellular proteinaceous toxins.

Most pathogens produce pectinases but *E. amylovora* appears to be an exception (Seemüller & Beer 1976) so they may not be essential for pathogenicity except in the soft rotting bacteria where they are a primary determinant; Wood (1978) suggested that early pectinase activity might induce plant defence mechanisms which would discourage establishment, and there appears to be no indication that they are important in the very early stages of infection.

Of the low molecular weight toxins and growth regulators, none appears essential in the early stages of infection though they could be contributory factors. (For further discussion of these agents see Kelman, 1979, and Ch. 5, this volume.)

(vii) *Plasmids*

The presence of plasmids has been demonstrated in a variety of plant pathogens but only rarely has an association with pathogenic potential been demonstrated (Chatterjee 1978; Chatterjee & Starr 1980). The most notable association is that of the Ti plasmid in *A. tumefaciens* which induces crown gall but this type of association appears to be a special case (see Ch. 7, this volume).

7. Conclusions

In addition to physical barriers, the plant has several lines of defence against invading bacteria. The balance of nutrients and inhibitors at sites of entry and establishment may be bactericidal, bacteriostatic or inhibitory; invading bacteria may be agglutinated or bound to cell walls, and finally, incompatible pathogens may induce a hypersensitivity reaction, following which, further multiplication and progression is prevented.

With a compatible pathogen, motility and a chemotactic response may aid entry and young host tissue probably provides a benign or stimulatory medium for growth. Agglutination and binding are avoided and HR is not induced; instead, plant cell water and nutrients are exploited by a controlled process. Together these allow rapid multiplication and establishment at an early stage.

Present evidence suggests that extracellular polysaccharides (EPS) have an important role in countering plant defence mechanisms and ensuring persistence of water-soaking, but apparently only in the compatible host–pathogen system. The role of lipopolysaccharides and their equivalent in Gram positive bacteria is less clear and the importance of other cell components and products in the early stages of infection is also uncertain.

The nature of the HR inducing agent and the means by which compatible pathogens ensure a more controlled leakage of plant cell water and nutrients have yet to be determined. It is possible that they are different manifestations of a single agent.

8. References

ALBERSHEIM, P. & ANDERSON-PROUTY, A. J. 1975 Carbohydrates, proteins, cell surfaces and the biochemistry of pathogenesis. *Annual Review of Plant Physiology* **26**, 31–52.

ANDERSON, A. J. & JASALAVICH, C. 1979 Agglutination of pseudomonad cells by plant products. *Physiological Plant Pathology* **15**, 149–159.

ANGADI, C. V. 1978 Extra-cellular slime of *Xanthomonas oryzae* in bacterial leaf blight of rice. *Phytopathologische Zeitschrift* **93**, 170–180.

AYERS, A. R., AYERS, S. B. & GOODMAN, R. N. 1979 Extracellular polysaccharide of *Erwinia amylovora*: a correlation with virulence. *Applied and Environmental Microbiology* **38**, 659–666.

AYRES, P. G. 1978 Water relations of diseased plants. In *Water Deficits and Plant Growth* Vol. V, ed. Kozlowski, T. T. pp.1–60. London & New York: Academic Press.

BATEMAN, D. F. & BASHAM, H. G. 1976 Degradation of plant cell walls and membranes by microbial enzymes. In *Physiological Plant Pathology* ed. Heitefuss, R. & Williams, P. H. pp.316–355. Berlin: Springer-Verlag.

BENNETT, R. A. & BILLING, E. 1978 Capsulation and virulence in *Erwinia amylovora*. *Annals of Applied Biology* **89**, 41–45.

BENNETT, R. A. & BILLING, E. 1980 Origin of the polysaccharide component of ooze from plants infected with *Erwinia amylovora*. *Journal of General Microbiology* **116**, 341–349.

BILLING, E. 1960 An association between capsulation and phage sensitivity in *Erwinia amylovora*. *Nature, London* **186**, 819–820.
BILLING, E. 1976 The taxonomy of bacteria on the aerial parts of plants. In *Microbiology of Aerial Plant Surfaces* ed. Dickinson, C. H. & Preece, T. R. pp.223–273. London & New York: Academic Press.
BILLING, E. & GARRETT, C. M. E. 1980 Phages in the identification of plant pathogenic bacteria. In *Microbial Classification and Identification* ed. Goodfellow, M. & Board, R. G. pp.319–338. Society for Applied Bacteriology Symposium Series No. 8. London & New York: Academic Press.
BRUEGGER, B. B. & KEEN, N. T. 1979 Specific elicitors of glyceollin accumulation in the *Pseudomonas glycinea*-soybean host-parasite system. *Physiological Plant Pathology* **15**, 43–51.
BUCHANAN, R. E. & GIBBONS, N. E. eds 1974 *Bergey's Manual of Determinative Bacteriology* 8th edn. Baltimore: Williams & Wilkins.
CALLOW, J. A. 1977 Recognition, resistance and the role of plant lectins in host–parasite interactions. *Advances in Botanical Research* **4**, 1–49.
CHATTERJEE, A. K. 1978 Genetics of phytopathogenic bacteria. In *Proceedings 4th International Conference on Phytopathogenic Bacteria* Angers, 1978, pp.3–16.
CHATTERJEE, A. K. & STARR, M. P. (1980) Genetics of *Erwinia* species. *Annual Review of Microbiology* **34**, 645–676.
CROSSE, J. E. 1956 Bacterial canker of stone fruits. II. Leaf scar infection of cherry. *Journal of Horticultural Science* **31**, 212–224.
CROSSE, J. E., GOODMAN, R. N. & SHAFFER, W. H. 1972 Leaf damage as a predisposing factor in the infection of apple shoots by *Erwinia amylovora*. *Phytopathology* **62**, 176–182.
DAFT, G. C. & LEBEN, C. 1972 Bacterial blight of soybean: epidemiology of blight outbreaks. *Phytopathology* **62**, 57–62.
DE VAY, J. E. & ADLER, H. E. 1976 Antigens common to hosts and parasites. *Annual Review of Microbiology* **30**, 147–168.
DIXON, R. A. & LAMB, C. J. 1980 The specificity of plant defences. *Nature, London* **283**, 135–136.
DOKE, N., GARAS, N. A. & KUĆ, J. 1979 Partial characterization and aspects of the mode of action of a hypersensitivity-inhibiting factor (HIF) isolated from *Phytophthora infestans*. *Physiological Plant Pathology* **15**, 127–140.
DYE, D. W., BRADBURY, J. F., GOTO, M., HAYWARD, A. C. & LELLIOTT, R. A. 1980 International standards for naming pathovars of phytopathogenic bacteria and a list of pathovar names and pathotype strains. *Review of Plant Pathology* **59**, 153–168.
EDEN-GREEN, S. J. & KNEE, M. 1974 Bacterial polysaccharide and sorbitol in fireblight exudate. *Journal of General Microbiology* **81**, 509–512.
EL-BANOBY, F. E. & RUDOLPH, K. 1979 Induction of water-soaking in plant leaves by extracellular polysaccharides from phytopathogenic pseudomonads and xanthomonads. *Physiological Plant Pathology* **15**, 341–349.
ELLWOOD, D. C., MELLING, J. & RUTTER, P. eds 1979 *Adhesion of Microorganisms to Surfaces*. London & New York: Academic Press.
ERCOLANI, G. L. 1973 Two hypotheses on the aetiology of response of plants to phytopathogenic bacteria. *Journal of General Microbiology* **75**, 83–95.
ERCOLANI, G. L. & CROSSE, J. E. 1966 The growth of *Pseudomonas phaseolicola* and related plant pathogens *in vivo*. *Journal of General Microbiology* **45**, 429–439.
ESSENBERG, M., CASON, E. T., HAMILTON, B., BRINKERHOFF, L. A., GHOLSON, R. K. & RICHARDSON, P. E. 1979*a* Single cell colonies of *Xanthomonas malvacearum* in susceptible and immune cotton leaves and the local resistant response to colonies in immune leaves. *Physiological Plant Pathology* **15**, 53–68.
ESSENBERG, M., HAMILTON, B., CASON, E. T., BRINKERHOFF, L. A., GHOLSON, R. K. & RICHARDSON, P. E. 1979*b* Localized bacteriostasis indicated by water dispersal of colonies of *Xanthomonas malvacearum* within immune cotton leaves. *Physiological Plant Pathology* **14**, 69–78.
FELICIANO, A. & DAINES, R. H. 1970 Factors influencing ingress of *Xanthomonas pruni*

through peach leaf scars and subsequent development of spring cankers. *Phytopathology* **60**, 1720–1726.

Feng, T-Y. & Kuo, T-T. 1975 Bacterial leaf blight of rice plant. VI. Chemotactic responses of *Xanthomonas oryzae* to water droplets exuded from water pores of the leaf of rice plants. *Botanical Bulletin of Academia Sinicia* **16**, 126–136.

Goodman, R. N. 1976 Physiological and cytological aspects of the bacterial infection process. In *Physiological Plant Pathology* Vol. 4 ed. Heitefuss, R. & Williams, P. H. pp.172–196. Berlin: Springer-Verlag.

Goodman, R. N., Huang, J. S. & Huang, P. Y. 1974 Host-specific phytotoxic polysaccharide from apple tissue infected by *Erwinia amylovora*. *Science, N. Y.* **183**, 1081–1082.

Goodman, R. N., Huang, P. Y., Huang, J. S. & Thaipanich, V. 1976*a* Induced resistance to bacteria. In *Biochemistry and Cytology of Plant-parasite Interaction* ed. Tomiyama, K., Daly, J. M., Uritani, I., Oku, H. & Ouchi, S. pp.35–42, Tokyo: Kadansha.

Goodman, R. N., Huang, P. Y. & White, J. A. 1976*b* Ultrastructural evidence for immobilization of an incompatible bacterium, *Pseudomonas pisi*, in tobacco leaf tissue. *Phytopathology* **66**, 754–764.

Hignett, R. C. & Quirk, A. V. 1979 Properties of phytotoxic cell-wall components of plant pathogenic pseudomonads. *Journal of General Microbiology* **110**, 77–81.

Hildebrand, D. C., Alosi, M. C. & Schroth, M. N. 1980 Physical entrapment of pseudomonads in bean leaves by films formed at air-water interfaces. *Phytopathology* **70**, 98–109.

Hildebrand, E. M. 1937 The blossom-blight phase of fireblight and methods of control. *Cornell University Agricultural Experiment Station, Memoir* **207**.

Hodgson, R., Peterson, W. H. & Riker, A. 1949 The toxicity of polysaccharides and other large molecules to tomato cuttings. *Phytopathology* **39**, 47–62.

Horino, O. 1976 Induction of bacterial leaf blight resistance by incompatible strains of *Xanthomonas oryzae*. In *Biochemistry and Cytology of Plant Parasite Interactions* ed. Tomiyama, K., Daly, J. M., Uritani, I., Oku, H. & Ouchi, S. pp.43–55. Tokyo: Kadansha.

Hsu, S-T. & Goodman, R. N. 1978 Agglutinating activity in apple cell suspension cultures inoculated with a virulent strain of *Erwinia amylovora*. *Phytopathology* **68**, 355–360.

Huang, P. Y. & Goodman, R. N. 1976 Ultrastructural modifications in apple stems induced by *Erwinia amylovora* and the fireblight toxin. *Phytopathology* **66**, 269–276.

Huang, J. S. & van Dyke, C. G. 1978 Interactions of tobacco callus tissue with *Pseudomonas tabaci*, *P. pisi* and *P. fluorescens*. *Physiological Plant Pathology* **13**, 65–72.

Husain, A. & Kelman, A. 1958 Relation of slime production to mechanisms of wilting and pathogenicity of *Pseudomonas solanacearum*. *Phytopathology* **48**, 155–165.

Inouye, M. Ed. 1979 *Bacterial Outer Membranes*. New York: John Wiley.

Johnson, J. 1947 Water-congestion in plants in relation to disease. *University of Wisconsin Research Bulletin* **160**, 1–35.

Kelman, A. 1979 How bacteria induce disease. In *Plant Disease* Vol. 4, ed. Horsfall, J. G. & Cowling, E. B. pp.181–202. London & New York: Academic Press.

Kelman, A. & Sequeira, L. 1972 Resistance in plants to bacteria. *Proceedings of the Royal Society B* **181**, 247–266.

Kuć, J. A. 1976 Phytoalexins. In *Physiological Plant Pathology* ed. Heitefuss, R. & Williams, P. H. pp.632–652. Berlin: Springer-Verlag.

Kuć, J. A. 1979 Modes of metabolic determination of specificity. In *Recognition and Specificity in Plant Host-Parasite Interactions* ed. Daly, J. M. & Uritani, I. pp.33–47. Tokyo: Japan Scientific Societies Press.

Leach, J. G. 1940 *Insect Transmission of Plant Diseases*. New York: McGraw-Hill.

Lelliott, R. A., Billing, E. & Hayward, A. C. 1966 A determinative scheme for the fluorescent plant pathogenic pseudomonads. *Journal of Applied Bacteriology* **29**, 470–489.

Lippincott, J. A. & Lippincott, B. B. 1977 Nature and specificity of the bacterium-host attachment in *Agrobacterium* infection. In *Cell Wall Biochemistry Related to Specificity in Host-Plant Pathogen Interactions* ed. Solheim, B. & Raa, J. pp.439–451. Oslo: Universitets forlaget.

LUCAS, L. T . & GROGAN, R. G. 1969 Some properties of specific antigens of *Pseudomonas lachrymans* and other *Pseudomonas* nomenspecies. *Phytopathology* **59**, 1913–1917.

LUND, B. M. 1979 Bacterial soft-rot of potatoes. In *Plant Pathogens* ed. Lovelock, D. W. pp.19–49. Society for Applied Biology Technical Series No. 12. London & New York: Academic Press.

MATTHEE, F. N. & DAINES, R. H. 1968 Effects of soil type and substrate aeration on stomatal activity, water diffusion pressure deficit, water congestion and bacterial infection of peach and pepper foliage. *Phytopathology* **58**, 1298–1301.

MAZZUCHI, U., BAZZI, C. & PUPILLO, P. 1979 The inhibition of susceptible and hypersensitive reactions by protein-lipopolysaccharide complexes from phytopathogenic pseudomonads: relationship to polysaccharide antigenic determinants. *Physiological Plant Pathology* **14**, 19–30.

MILLER, P. W. 1929 Studies of fireblight of apple in Wisconsin. *Journal of Agricultural Research* **39**, 579–621.

PANAGOPOULOS, C. G. & CROSSE, J. E. 1964 Blossom blight and related symptoms caused by *Pseudomonas syringae* van Hall on pear trees. *Annual Report East Malling Research Station* for 1963, pp.119–122.

PANOPOULOS, N. J. & SCHROTH, M. N. 1974 Role of flagellar motility in the invasion of bean leaves by *Pseudomonas phaseolicola*. *Phytopathology* **64**, 1389–1397.

PITCHER, R. S. 1963 Role of plant-parasitic nematodes in bacterial diseases. *Phytopathology* **53**, 35–39.

POWELL, D. A. 1979 Structure, solution properties and biological interactions of some microbial extracellular polysaccharides. In *Microbial Polysaccharides and Polysaccharases* ed. Berkeley, R. C. W., Gooday, G. W. & Ellwood, D. C. pp.117–160. London & New York: Academic Press.

QUIRK, A. V., SLETTEN, A. & HIGNETT, R. C. 1976 Properties of phage-receptor lipopolysaccharide from *Pseudomonas morsprunorum*. *Journal of General Microbiology* **96**, 375–381.

SANDS, D. & ZUCKER, M. 1976 Amino acid inhibition of pseudomonads and its reversal by biosynthetically related amino acids. *Physiological Plant Pathology* **9**, 127–133.

SASSER, M. 1978 Involvement of bacterial protein synthesis in induction of the hypersensitivity reaction in tobacco. *Phytopathology* **68**, 361–363.

SCHMIT, J. 1978 Microscopic study of early-stages of infection by *Pseudomonas solanacearum* E.F.S. on *in vitro* grown tomato seedlings. *Proceedings 4th International Conference on Plant Pathogenic Bacteria* Angers, 1978, pp.841–856.

SCHÖNBECK, F. & SCHLÖSSER, E. 1976 Preformed substances as potential protectants. In *Physiological Plant Pathology* ed. Heitefuss, R. & Williams, P. H. pp.653–678. Berlin, Heidelberg, New York: Springer-Verlag.

SEEMÜLLER, E. A. & BEER, S. V. 1976 Absence of cell wall polysaccharide degradation by *Erwinia amylovora*. *Phytopathology* **66**, 433–436.

SEQUEIRA, L. 1976 Induction and suppression of the hypersensitive reaction caused by phytopathogenic bacteria: specific and non-specific components. In *Specificity in Plant Diseases* ed. Wood, R. K. S. & Granati, A. pp.289–306. New York & London: Plenum Press.

SEQUEIRA, L. 1978 Lectins and their role in host-pathogen specificity. *Annual Review of Phytopathology* **16**, 453–481.

SEQUEIRA, L. 1979 The acquisition of systemic resistance by prior inoculation. In *Recognition and Specificity in Plant Host-Parasite Interactions* ed. Daly, J. M. & Uritani, I. pp.231–251. Tokyo: Japan Scientific Societies Press.

SEQUEIRA, L. & GRAHAM, T. L. 1977 Agglutination of avirulent strains of *Pseudomonas solanacearum* by potato lectin. *Physiological Plant Pathology* **11**, 43–54.

SEQUEIRA, L., GAARD, G. & DE ZOETEN, G. A. 1977 Interaction of bacteria and host cell walls: its relation to mechanisms of induced resistance. *Physiological Plant Pathology* **10**, 43–50.

SJULIN, T. M. & BEER, S. V. 1978 Mechanism of wilt induction by amylovorin in cotoneaster shoots and its relation to wilting of shoots infected by *Erwinia amylovora*. *Phytopathology* **68**, 89–94.

STALL, R. E. & COOK, A. A. 1979 Evidence that bacterial contact with the plant cell is necessary for the hypersensitivity reaction but not the susceptible reaction. *Physiological Plant Pathology* **14**, 77–84.

STAUB, T. & WILLIAMS, P. H. 1972 Factors influencing black rot lesion development in resistant and susceptible cabbage. *Phytopathology* **62**, 722–728.

STROBEL, G. A. 1977 Bacterial phytotoxins. *Annual Review of Microbiology* **31**, 205–224.

STROBEL, G. A. & HESS, W. M. 1968 Biological activity of a phytotoxic glycopeptide produced by *Corynebacterium sepedonicum*. *Plant Physiology* **43**, 1673–1688.

SUTHERLAND, I. ed. 1977 *Surface Carbohydrates of the Prokaryotic Cell.* London & New York: Academic Press.

SUTTON, J. C. & WILLIAMS, P. H. 1970 Comparison of extracellular polysaccharide of *Xanthomonas campestris* from culture and from infected cabbage leaves. *Canadian Journal of Botany* **48**, 645–651.

TABEI, H. 1973 Infection mechanism of *Xanthomonas oryzae* to rice plant, with special reference to the mode of bacterial invasion associated with the wilt symptom. *Shokubutsu Byogai Kenkyu* **8**, 191–202.

THOMSON, S. V. 1978 Stigmatic surfaces of pear flower pistils as a source of inoculum for *Erwinia amylovora*. *Proceedings 4th International Conference on Plant Pathogenic Bacteria* Angers, 1978, p.816.

TRIGALET, A., SAMSON, R. & COLÉNO, A. 1978 Problems related to the use of serology in phytobacteriology. *Proceedings 4th International Conference on Plant Pathogenic Bacteria* Angers, 1978, pp.271–288.

VAN ALFEN, N. K. & ALLARD-TURNER, V. 1979 Susceptibility of plants to vascular disruption by macromolecules. *Plant Physiology* **63**, 1072–1075.

VANDERPLANK, J. E. 1978 *Genetic and Molecular Basis of Plant Pathogenesis.* Berlin: Springer-Verlag.

WALLIS, F. W. & TRUTER, S. J. 1978 Histopathology of tomato plants infected with *Pseudomonas solanacearum*, with emphasis on ultrastructure. *Physiological Plant Pathology* **13**, 307–317.

WOOD, R. K. S. 1978 Enzymes produced by fungi and bacteria. *Annales de Phytopathologie* **10**, 127–135.

YOUNG, J. M. 1974*a* Development of bacterial populations *in vivo* in relation to plant pathogenicity. *New Zealand Journal of Agricultural Research* **17**, 105–113.

YOUNG, J. M. 1974*b* Effect of water on bacterial multiplication in plant tissue. *New Zealand Journal of Agricultural Research* **17**, 115–119.

The Progression of Bacterial Disease Within Plants

T. F. PREECE

Department of Plant Sciences,
The University of Leeds, Leeds, UK

Contents

1. Introduction

PLANT PATHOGENIC BACTERIA can sometimes enter plants through natural openings, but the significance, in terms of infection, of the bacteria that enter this way, as compared with entry via wounds, cannot yet be assessed: more evidence is needed. It is the present consensus of opinion however (see Ch. 4, this volume) that wounds are of major importance for the entry of plant pathogenic bacteria into plants. For entry to occur via a wound the pathogenic bacteria will have to be available at the cut surface, e.g. on an infected cutting knife or in plant debris containing the pathogen, during the time interval between wounding and sealing of the wound by tissues arising from the usual wound cambium that develops. Two properties of the pathogen, basic to progression of the disease and apparently antithetical, may play a part at the cut surface. First, bacteria may adhere strongly to the naked cut surface, probably much as *Agrobacterium tumefaciens* has been shown to do to potato tissue slices (Glogowski & Galsky 1978). It seems likely, because of results from the work with *Agrobacterium* and from quantitative inoculation of entire, intact leaves, that pathogens will attach much more strongly than non-pathogens (Preece & Wong 1981). The bacterial plant pathogen will thus be placed, by these attachment processes, in the wound amongst the cells of the host tissue, including the intercellular areas, and in vascular areas, including the large xylem vessels. Plant pathogenic bacteria are characteristically found in large numbers later, on and deeper in the tissues of these two zones—the intercellular spaces and the xylem vessels. Secondly, and apparently opposite in function to this likely attachment process in an anatomically advantageous position, there is a process of ‘non-binding’ of pathogens in plants, as distinct from ‘binding’ of saprophytes and avirulent forms (Ch. 4, this volume) whereby pathogens remain free to move in the plant. However, the significance of these two phenomena may be resolved by future research. Direct entry of pathogenic bacteria into xylem vessels after wounding (e.g. of *Corynebacterium michiganense*, Pine *et al.* 1955) takes place in many diseases. At varying intervals of time after wounding, perhaps years later (*Erwinia salicis* in willow wood; Adegeye & Preece 1978) or minutes later (*Xanthomonas pelargonii* in pelargonium cuttings, McPherson & Preece 1979) the bacterial pathogen may be seen within the vascular tissues, before any macroscopic symptoms are noticeable. This systemic invasion, at first symptomless, is a central practical problem for growers of crops that are propagated vegetatively. Cuttings, of pelargonium infected with *Xanthomonas pelargonii* (McPherson & Preece 1979), and of cricket bat willow containing *Erwinia salicis*; or tubers (anatomically equivalent to stems) of potato infected with *E. carotovora* var. *atroseptica* (Lapwood & Hide 1971) are examples of such systemic infections.

Bacteria in protected positions on plants (within buds; within seed coats) are probably correctly interpreted as being on plants rather than in them, but as with infected propagating material, seed infections are of great importance in practice and difficult to control. Sometimes the cotyledons may be infected, but whether this is so or not the disease ‘progresses’ as the seed germinates. The list of diseases caused by bacteria which, in certain instances, have been shown to be seed-borne, is very long. Halo-blight of beans, caused by *Pseudomonas phaseolicola*, black-arm of cotton caused by *Xanthomonas malvacearum* and bacterial wilt disease of tomato caused by *Corynebacterium michiganense*, are examples.

Plant material, including entire plants, is moved about a great deal, even between countries, and it must be borne in mind that an ‘outbreak’ of infection, for example, of *Xanthomonas begoniae* in a begonia crop, may not have arisen by the bacteria ‘entering’ the plants in some way in the glasshouse; they may have been infected when they arrrived there as young plants, and disease development may occur if the environmental conditions are suitable.

The brief account which follows owes much to the work of students who have worked with me on bacterial diseases of willow, begonia, pelargonium, ash, ryegrass, tomato and mushroom, and it is a partial account. There are probably 250,000 distinct species of plants (Curtis 1975) wild and cultivated; workers on bacterial plant diseases have been, and are, very few; and our ignorance in this area of applied plant bacteriology is conspicuous. From the poorly understood early stages of infections, one can now turn to the easily visible disease symptoms of bacterial plant diseases.

2. Symptoms of Bacterial Disease in Plants

In plants, diseases caused by bacteria, viruses, fungi, mycoplasmas and rickettsia-like organisms show a wide range of symptoms, the most common of which are: (i) colour change, (ii) wilting, (iii) distortion, (iv) gall formation, (v) vascular darkening, (vi) rotting, (vii) stunting. In addition to these, plants diseased by bacteria most commonly show: (i) 'water-soaking' (apparent waterlogging) of the tissues at lesion sites, (ii) angularity of necrotic areas on infected leaves, (iii) the production of bacterial 'ooze'.

If discoloration of the vascular tissues (also a symptom of fungal disease) accompanies these three distinctive symptoms, it is a most useful pointer to the probability of a plant disease, observed in the field, being bacterial in origin. Of course, it is necessary to exclude infection by fungi, viruses and mycoplasmas, and physiological defects, by laboratory examination and to isolate the bacterium and fulfill Koch's postulates before a confirmatory diagnosis is complete.

Water-soaking of bacterially infected tissue and the probable involvement of extracellular polysaccharide, is discussed in Ch. 4 of this volume. The angularity of leaf lesions seems to have been investigated very little anatomically; it is usually thought to be due to restriction of lateral movement of the bacteria through mesophyll tissue by the vascular tissue (veins) and their supporting external tissues. Ooze emerging from bacterially infected plants comprises bacteria and their associated extracellular materials.

Discoloration of the tissues of plants after invasion by bacteria is generally associated with changes in the phenolic constituents of the cells. In the discoloration which is the main symptom of watermark disease of the cricket bat willow caused by *E. salicis*, these changes are complex: the symptom of xylem darkening is associated with changes in pectic materials, cellulosic substances, lignin and polyphenols (Wong & Preece 1978*a*,*b*,*c*). The overall picture in bacterially infected willow wood is of a succession of reactions and inter-reactions associated with the presence in the tissue of the bacterium *E. salicis*, the latter genetically endowed with factors which account for what we loosely call 'bacterial pathogenicity'.

Reddening, when it occurs in plants infected by bacteria, is due to changes in anthocyanins, pigments peculiar to plants. Polyphenol oxidase under acid conditions converts, e.g. colourless leucoanthocyanins to coloured anthocyanins (the brightly coloured compounds of many flowers and fruits). Coloured anthocyanins may be synthesized also from simple compounds in bacterially diseased tissue (Wong & Preece 1978*c*).

Yellowing, or chlorosis, another colour change, may be caused in a variety of ways around bacterial lesions on plants. Non-specific yellowing is generally caused by toxins. An example of these, phaseolotoxin in *Pseudomonas phaseolicola* infections of bean, is an inhibitor of orthinine carbamoyl transferase activity, and orthinine accumulates around the lesions. In fungal diseases, yellowing has long been associated with ethylene production (as in rose black spot caused by *Diplocarpon rosae*); ethylene also causes defoliation, without chlorosis, of citrus trees infected with *Xanthomonas*. Tabtoxin activity (involved in chlorosis of tobacco infected with *Pseudomonas tabaci*) leads to ammonia production in the tissues. Most cases of chlorotic symptoms have not yet been investigated, but degeneration of chloroplasts around lesions is common and is thought, in general, to be caused by the toxins produced by the pathogenic bacteria in the tissues.

Wilting often precedes other symptoms of bacterial plant diseases—growers often report wilting of pelargonium cuttings before *Xanthomonas pelargonii* infection is even suspected. Adegeye & Preece (1978) noted wilting after artificial inoculation of willow with *E. salicis*. Irreversible wilting may in fact be the principal symptom of a bacterial disease: rye grass wilt (caused by *Xanthomonas graminis*) is such a disease (Wilkins & Exley 1977). Wilt induction is usually associated with production of extracellular bacterial polysaccharide (Ch. 4, this volume) and the interaction of this material with the vascular tissues of the plant. Wilting needs further study in bacterial plant diseases–in some cases, such as *Corynbacterium insidiosum* wilt of lucerne (*Medicago sativa*), certain glycopeptide molecules act as an extracellular toxin.

Distortion of foliage is especially conspicuous when pathogens damage leaves when they are young and expanding; this effect is seen clearly in early infections of *Pseudomonas syringae* on *Philadelphus*.

Gall formation by *Agrobacterium tumefaciens* has recently been much studied and is now known to be the result of the transformation of plant cells by a plasmid carried by pathogenic strains of this bacterium (Ch. 7, this volume). In so-called ash 'canker', also a neoplastic plant disease (Boa & Preece 1979), the cause of stimulation of cells by *Pseudomonas savastanoi* has not been elucidated, but a very closely allied organism causing galls on oleander (Wilson 1965) is now known to produce indolyl 3-acetic acid (IAA) *in vitro* (Smidt & Kosuge 1978). Although the phenomona of hyperplasia (increased numbers of cells) and hypertrophy of individual cells may be seen in many fungal diseases (club root of brassicas, caused by *Plasmodiophora brassicae*; azalea gall caused by *Exobasidium azaleae*), the 'swellings' noted after many bacterial infections of plants need more study. The development of meristematic tissue, followed by swelling, in *Pseudomonas caryophylli* infections of carnation tissue (Nelson & Dickey 1970) has been noted as an important symptom. Recently McPherson (1981) recorded meristematic activity in tissues of pelargonium inoculated with an avirulent isolate of *Xanthomonas pelargonii*; this did not occur in the usual field outbreaks of the disease associated with virulent strains of the pathogen.

It is note-worthy that there are close similarities between insect induced galls and those induced by bacteria; the latter are however usually limited in their size, as are *Rhizobium* induced nodules (Ch. 3, this volume). The cause of many galls of trees and herbaceous plants is unknown at present.

Soft rotting of plant tissue, often rapid, may be caused by 'pectolytic' plant-pathogenic bacteria such as *E. carotovora* var. *carotovora* (Ch. 9, this volume). Pectic enzymes have a much wider significance, however, participating in the development of cytolytic or cytotoxic symptoms, leading ultimately to necrosis. Polygalacturonic acid *trans*-eliminase (PGTE) is involved in initial infection processes (Zucker & Hankin 1970). The complexity of the host–parasite interaction needs constant emphasis even in a generalized account such as this. The cytolytic or cytotoxic factor(s) involved in a particular situation may not be due to a pectolytic enzyme(s)—it may be due to a glycopeptide toxin as mentioned earlier—of which the toxin present in ring rot infections of potato (caused by *Corynebacterium insidiosum*) is another example. In some bacterial diseases, pectolytic enzymes seem not to be involved at all in pathogenesis; *E. amylovora* infection is a possible example of this.

In man, the immune system of antibody–antigen reactions involving the production and reactions of specific γ-globulins is of the greatest importance in disease control. This process does not occur in plants: plants have no immune system, nor do they have any mobile cells capable of engulfing and destroying bacteria. The 'chemical resistance' of plant tissues to bacteria (whether constitutive or induced) is not understood, and the enormous volume of work on phytoalexins (antimicrobial chemicals) produced by plants under a variety of stresses, and almost certainly concerned with the containment of fungal infections of plants, is not considered at present to be an over-riding phenomenon accounting for the resistance of plants to bacterial diseases.

3. Microscopy of Diseased Plant Tissues

Our quantitative conceptual models of what happens at the microscopic level during the progression of bacterial disease in plants are very imperfect at present.

The first person to study a number of bacterial plant diseases (and indeed to convince his contemporaries that they existed at all) was Erwin F. Smith (1920) who made some general comments that are still accepted. He considered that the active growth phases of plants are the ones during which the greatest spread of bacteria occurs within the plant. Progression of the disease bacteria is time-limited because the ‘hardening’ of plant tissues restricts the movement and activity of the pathogen. Immature tissues are most susceptible, and ‘abundant juiciness’ favours the whole disease development process most of all. The limitation of growth in mature tissues is still not understood and technical problems control our studies of the histology of diseased plant tissues (Preece 1981). Bacterial numbers have to be high in order to locate them at all, using an oil immersion objective. Very large numbers of bacteria in tissues are probably essential before we can conclude much from either light or electron microscopy (e.m.). It is notoriously difficult to find bacteria in e.m. sections of even heavily inoculated plants. Thus we know little, cytologically, about the earliest stages of infection of plants by bacteria for technical reasons.

Bacteria can be found in wound infections of *Xanthomonas pelargonii* (McPherson 1981) when quite massive numbers are present in the intercellular spaces wherein the bacteria multiply and carry out their biosynthetic and catabolic activities. Much will depend on the virulence of the pathogen. The next step in the spread of the pathogen involves, almost universally, entry into the vascular tissue (e.g. *E. amylovora*, Hockenhull 1979). When infected vascular tissue of a leaf, for example, is cut through, bacteria can be seen pouring out into the water in which the tissue is mounted for microscopic observation. No one yet knows exactly how the vascular tissue is entered by bacteria from the surrounding parenchyma (Nelson & Dickey 1970): as with the infection of plants by fungi the processes are likely to be either enzymatic (e.g. after cellulolytic dissolution of cell walls) or physical (after the rupture of cell walls by pressure, especially rupture of the more delicate pit membranes of xylem vessels). It is not yet clearly established whether bacteria move at all of their own volition in plants or whether they are moved by the normal processes occurring in vascular tissue (Section 4). Just as we do not know how bacteria get into the differentiated vascular tissues we are equally ignorant as to how they break out of them. Bacteria can be seen passing out between the spiral thickenings in the vessels of tomato stems infected by *Corynebacterium michiganense*, from whence they move out into the parenchyma. In some histological studies 'pockets' of bacteria in cortical and other tissues have been reported, e.g. in the xylem parenchyma of carnations affected by *Ps. caryophylli*; although not precisely analogous, the limited lesion of *Ps. savastanoi* in ash canker contains 'pockets' of bacteria and corky tissue (Boa 1981). These complex host–parasite interaction areas are neither locatable accurately nor understood, but they are probably important epidemiologically. The bacteria appear to be tolerated as saprophytes and probably survive much longer in this way. More usually, in leaves, massive multiplication in the intercellular spaces merely results in necrosis. Attempts to summarize the phases seen in microscopical studies of diseased plants are presented in Fig. 1.

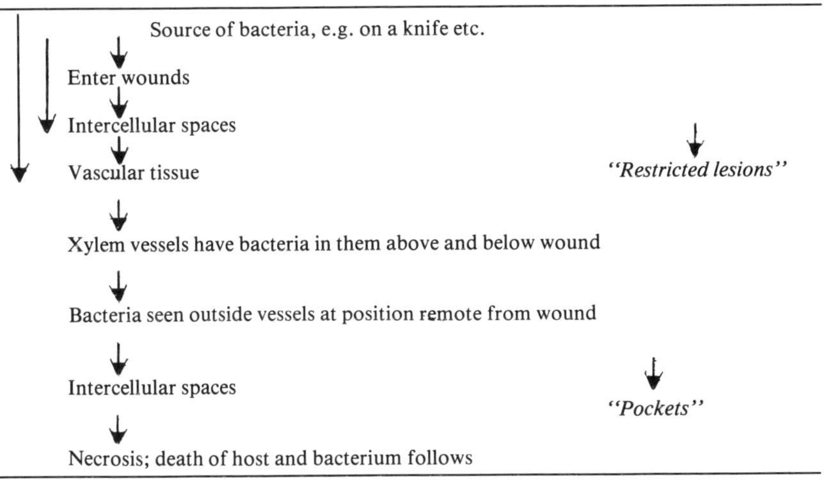

Fig. 1. A diagramatic representation of the microscopical phases of many bacterial diseases in plants, which may or may not be accompanied by visible symptoms. Very little is known about most of these phases, even in qualitative terms. The time scale varies with the disease and with the extent of infection of seed or cuttings that may become sources of disease in epidemics.

4. Numbers of Bacteria Involved

Single bacterial cells may initiate infections (Ch. 4, this volume) but before any of the symptoms discussed in Section 2 (above) appear, it is most usual that a threshold high bacterial population is produced within the plant (Erinle 1975). This is apparently so in black-leg of potato caused by *E. carotovora* var. *atroseptica*.

A comparison of quantitative studies on *E. salicis* infections (Adegeye & Preece 1978) in a tree, *Xanthomonas perlargonii* (McPherson & Preece 1979) in a succulent stem, and *Pseudomonas tolaasii* in mushrooms (Wong & Preece 1981), is interesting.

Throughout experiments with 1·5-, 2·5- and 4·5-year-old small trees of *Salix alba* var. *caerulea*, *E. salicis* was recovered in the highest numbers from 1 cm lengths of the main stem which included the incoculation site itself. Whether symptoms developed or not (and they did not do so in half the young trees within the six months after inoculation) the number of cells of *E. salicis* in the trees increased dramatically (to $1{\cdot}05 \times 10^{10}$ per total of 4×1 cm lengths tested) up to one week after inoculation, but declined equally steeply (after two weeks to 3×10^{9} cells) with later isolations, e.g. at four weeks, indicating a persistent low level ($< 5 \times 10^{8}$ cells). The 'classical' internal symptom, dark brown internal wood staining, did not occur until four weeks after inoculation. This decline in numbers has also been reported in *E. amylovora* infections (Schroth *et al.* 1974) in these words ". . . the population of bacteria (after inoculation) subsequently declined markedly . . . the survival of *E. amylovora* is best in or near (the originally) infected tissues . . . host defence mechanisms 'purge' the bacteria from the tissues . . .". We have no understanding of these mechanisms.

In a succulent stem, the pathogen *Xanthomonas pelargonii* apparently moved very rapidly indeed, the more so downwards, in inoculated stem cuttings: 5 cm of the length of a 7 cm cutting contained the bacterium within 15 min. In less than a week bacterial numbers were very high throughout the cuttings. Water-soaking symptoms were seen after 5 days ($\pm 10^{8}$ bacteria/g wet weight) and necrosis after 26 days, when the population was somewhat higher. As in willow wood, the highest numbers were found persistently at the inoculation point, and numbers declined the further the sample was taken up or down the cutting from this site. In the case of *X. pelargonii*, the pathogen did not decline in numbers in the symptomless tissue as in *E. salicis* infections of willow wood. The *X. pelargonii* clearly overcame defence mechanisms of the plant and proceeded systemically throughout the cuttings. The pathogen must, as noted in Section 3 (above), have escaped from the xylem because visible symptoms (stem rot, the main symptom of the disease in the field) only occurred when the pathogen invaded the parenchymatous tissues. Symptom production was associated with peak numbers of bacteria in the pelargonium tissues.

The lack of structural complexity of the larger fungi, compared with a willow tree or pelargonium plant, allows experimental determination of inoculum thresholds for symptom production in a way that is not possible in other host–bacterium studies. Droplets of water containing known suspensions of, for example, *Ps. tolaasii* cells, can be used as inoculum on the caps of mushrooms. The number of cells of *Ps. tolaasii* essential for symptom production on a 6 cm diameter mushroom cap is $5{\cdot}4 \times 10^6$, and numbers below this result in symptomless infections. Under the normal conditions of mushroom cultivation these levels of bacterial population are reached rapidly (within five days of the inoculation of the compost with mycelium) and numbers increase logarithmically as the mushrooms grow. The picture here is very clear: bacterial numbers increase on and in the host, and symptoms appear dramatically as soon as a definable inoculum threshold is passed.

Thus in these examples (and there have been relatively few quantitative studies of this type, other than those cited by Billing (Ch. 4, this volume)), the results indicate that there are great variations in the numbers of bacteria present at different times after inoculation. High numbers of bacteria appeared to be necessary for symptom development in the above-mentioned three diseases. Quantitative studies are needed for almost every bacterial plant disease, and so generalization seems very unwise at the present state of knowledge. Recent work on ash canker, a neoplastic disease caused by a strain of *Ps. savastanoi* (Boa & Preece 1979) seems to indicate symptom production with very low numbers of bacteria in the tissues, the possibility being that this prominent disfigurement of trees is 'incited' by the causative organism in some way similar to the galls caused by *Agrobacterium tumefaciens*.

5. The Extent of Bacterial Infection within Plants

Diseases which destroy whole plants are relatively unusual. Complete destruction of plants, as for potato plants with fungal blight infections of the foliage caused by *Phytophthora infestans*, is approached however in the destruction of potato tubers by *E. carotovora* var. *carotovora*, although stem and leaf lesions are generally restricted in some way. Branch die-back and other symptoms are seen as in *E. salicis* infections (Preece 1977) whereby the plant (willow) dies eventually from a complex of physiological and anatomical changes initiated by the bacterial infection. This process may take many years in a tree. Gross disfigurement or a commercially unacceptable plant is the usual outcome of a bacterial disease. Whether symptoms develop or not, the bacteria may remain in the plant tissues after infection and the question arises as to the extent of this symptomless infection and how rapidly it may develop. This aspect is of enormous practical importance in relation to the production of plant cuttings for use as propagating material.

In experimental leaf petiole inoculations of *Corynebacterium michiganense* into tomato stems, Pine *et al.* (1955) observed bacteria in and around the xylem vessels five days after inoculation, 13·2 mm above the point of inoculation and 32 mm below it, the bacteria being mainly confined to the spiral xylem vessel elements. 'Pocket' formation near the vessels developed later during the 18 days over which movement was studied. Thus the bacterium moved 4 or 5 cm within the stem in 5 days. As the plants were '5 week old seedlings' and the length of the stem was not given, the results are difficult to compare precisely with those for *Xanthomonas pelargonii* in pelargonium cuttings (McPherson & Preece 1979). The speed of spread of the bacteria in pelargonium (5 cm in a 7 cm tall cutting in 15 min) was apparently more rapid than that of *C. michiganense* in tomato, for Pine *et al.* (1955) noted that the latter organism had moved only 2·33 mm from the inoculation site in 15 min. More detailed comparative studies are necessary.

In woody shoots of apple, Gowda & Goodman (1970) noted that *E. amylovora* moved 15·25 cm in 7 h, and 70 cm in 14 days. Apparently, this is a much greater movement than was observed (Adegeye & Preece 1978) in willow wood inoculated with *E. salicis*, where the inoculum moved at a slow rate. In young trees the movement of *E. salicis* was restricted in our experiments to an area near to the inoculation site (4–8 cm above and 4–16 cm below) in 30–60 cm long stems. The detailed evidence (Adegeye & Preece 1978) indicate that the bacterium appeared to remain in the area shown to be colonized at the first time of study (3 h after inoculation) for up to 6 months, when the experiment was terminated.

As in a consideration of the numbers of bacteria involved in the infection of plants, the extent of bacterial invasion of the tissues, probably always at first symptomless, varies greatly for different diseases. More work with marker saprophytes and inert particles the same size as bacteria, such as that attempted by Beckman & Keller (1977) in their study of the movement of fungal spores in xylem vessels is needed. The precise involvement of anatomical features (vessel ends) and physiological (transpiration stream) effects in plants makes this a challenging topic for research.

6. Bacteria in Plant Senescence and Decay

The epidemiological importance of the survival of plant pathogens after the decay and death of the plant is so great that this itself is a topic worthy of a separate review although for the sake of completeness some matters are raised here. What happens as the plant ages and dies? There appears to have been no detailed qualitative or quantitative study of bacteria in or on plants during senescence. The bacteria may of course 'survive' in herbaceous or woody cuttings (Section 2) taken before the parent or stock plant dies (e.g. carnations, pelargoniums, willow). The progression of the disease may result in a vital part of the plant (seed, bulb, corm or tuber used for vegetative propagation) becoming infected. In the case of *Pseudomonas phaseolicola* (Dowson 1957) the bacteria may pass along the vascular tissue of the funicle. The inner testa and outer layers of the cotyledons of seeds may be infected. The same author reports that *Xanthomonas malvacearum* enters seeds and survives as shiny masses in and around the micropyle, and that *Corynebacterium michiganense* plugs the micropyle and surrounds the embryo.

Carry-over of bacteria in the field seems important only if the organisms in the infected leaf material or other plant debris survive until another susceptible crop is available. Dowson (1957) also recorded that *C. michiganense* will survive in soil containing diseased plant material, and that 50% of plants raised in such soil may become infected. The survival of plant pathogenic bacteria for long periods in the absence of plant tissue is emphasized by many workers as being unlikely: *Agrobacterium tumefaciens* seems an exception to this rule. Recent findings (Wong & Preece 1980) that the sources of *Ps. tolaasii* on mushroom farms are the limestone and the peat brought on to the farms introduces another practical problem.

Survival in decaying plant material may be unexpectedly long. After the felling of willow trees infected with *Erwinia salicis*, Wong *et al.* (1974) reported the isolation of the pathogen five years later from grossly decayed stumps. On the other hand, the isolation of *Xanthomonas pelargonii* from the stem rotted bases of infected pelargonium cuttings (McPherson & Preece 1979) rapidly became impossible because of invading soil saprophytes. This was also the case in *Xanthomonas begoniae* infections (Taylor 1980), where some of the antagonistic organisms (*Bacillus* & *Pseudomonas* spp.) themselves showed promise in biological control experiments. Research in the field on the death of individual plant pathogenic bacteria in their decaying host plants is needed, and, again there appear to be few generalizations that are currently useful.

7. References

ADEGEYE, A. O. & PREECE, T. F. 1978 *Erwinia salicis* in cricket bat willows: rate of movement of the bacterium and the production of symptoms in young trees and shoots. *Journal of Applied Bacteriology* **44**, 265–277.

BECKMAN, C. H. & KELLER, J. L. 1977 Vessels do end! *Phytopathology* **67**, 954–956.

BOA, E. R. 1981 Ash canker disease. PhD Thesis, The University of Leeds, Leeds, UK.

BOA, E. R. & PREECE, T. F. 1979 *Pseudomonas* var. *fraxini*: symptoms of ash canker disease in the field in Scotland and England. *Proceedings Fourth International Conference on Plant Pathogenic Bacteria* Angers, 1978.

CURTIS, H. 1975 *Biology* 2nd edn. New York: Worth Publishers.

DOWSON, W. J. 1957 *Plant Diseases due to Bacteria*. Cambridge: Cambridge University Press.

ERINLE, L. D. 1975 Growth of *Erwinia carotovora* var. *atroseptica* and *E. carotovora* var. *carotovora* in potato stems. *Plant Pathology* **24**, 224–229.

GLOGOWSKY, W. & GALSKY, A. G. 1978 *Agrobacterium tumefaciens* site attachment as a necessary prerequisite for crown gall tumour formation on potato discs. *Plant Physiology* **61**, 1031–1033.

GOWDA, S. S. & GOODMAN, R. N. 1970 Movement and persistence of *Erwinia amylovora* in shoot, stem and root of apple. *Plant Disease Reporter* **54**, 576–582.

HOCKENHULL, J. 1979 *In situ* detection of *Erwinia amylovora* antigen in symptomless petiole and stem tissue by means of the fluorescent antibody technique. In *Yearbook (1979)* pp.1–14. Copenhagen: Royal Veterinary and Agricultural University.

LAPWOOD, D. H. & HIDE, G. A. 1971 Potato. In *Diseases of Crop Plants* ed. Western, J. H. London: Macmillan.

MCPHERSON, G. M. 1981 *Xanthomonas pelargonii* infections of *Pelargonium*. PhD Thesis, The University of Leeds, Leeds, UK.

MCPHERSON, G. M. & PREECE, T. F. 1979 Bacterial blight of *Pelargonium*: movement, symptom production and distribution of *Xanthomonas pelargonii* (Brown) Starr & Burkholder in *Pelargonium hortorum* Bailey, following artificial inoculation. *Proceedings Fourth International Conference on Plant Pathogenic Bacteria* Angers, 1978.

NELSON, P. E. & DICKEY, R. S. 1970 Histopathology of plants infected with vascular bacterial pathogens. *Annual Review of Phytopathology* **8**, 259–280.

PINE, T. S., GROGAN, R. G. & HEWITT, W. B. 1955 Pathological anatomy of bacterial canker of young tomato plants. *Phytopathology* **45**, 267–271.

PREECE, T. F. 1977 *Watermark Disease of the Cricket Bat Willow*. Forestry Commission Leaflet No. 20. London: Her Majesty's Stationery Office.

PREECE, T. F. 1981 Laboratory technology for plant pathologists, yesterday, today and tomorrow. *Review of Plant Pathology* **60**, 150–152.

PREECE, T. F. & WONG, W. C. 1981 Detectable attachment of bacteria to intact plant and fungal surfaces. In *Microbial Ecology of the Phylloplane* ed. Blakeman, J. P. pp.399–410. London & New York: Academic Press.

SCHROTH, M. N., THOMPSON, S. V., HILDEBRAND, D. C. & MOLLER, W. J. 1974 Epidemiology and control of fireblight. *Annual Review of Phytopathology* **12**, 389–412.

SMIDT, M. & KOSUGE, T. 1978 The role of indole-3-acetic acid accumulation by alpha methyl tryptophan resistant mutants of *Pseudomonas savastanoi* in gall formation on oleanders. *Physiological Plant Pathology* **13**, 203–214.

SMITH, E. F. 1920 *An Introduction to Bacterial Diseases of Plants*. Philadelphia & London: W. B. Saunders.

TAYLOR, E. H. 1980 Interactions with saprophytic bacteria in infection of *Begonia* by *Xanthomonas begoniae*. MPhil Thesis, The University of Leeds, Leeds, UK.

WILKINS, P. & EXLEY, J. K. 1977 Bacterial wilt of ryegrass in Britain. *Plant Pathology* **26**, 99.

WILSON, E. E. 1965 Pathological histogenesis in oleander tumors produced by *Pseudomonas savastanoi*. *Phytopathology* **55**, 1244–1249.

WONG, W. C. & PREECE, T. F. 1978*a* *Erwinia salicis* in cricket bat willows: histology and histochemistry of infected wood. *Physiological Plant Pathology* **12**, 321–332.

WONG, W. C. & PREECE, T. F. 1978*b* *Erwinia salicis* in cricket bat willows: peroxidase, polyphenoloxidase, β-glucosidase, pectinolytic and cellulolytic enzyme activity in diseased wood. *Physiological Plant Pathology* **12**, 333–347.

WONG, W. C. & PREECE, T. F. 1978*c* *Erwinia salicis* in cricket bat willows: phenolic constituents in healthy and diseased wood. *Physiological Plant Pathology* **12**, 349–357.

WONG, W. C. & PREECE, T. F. 1980 *Pseudomonas tolaasii* in mushroom crops: a note on the primary and secondary sources of the bacterium on a commercial farm in England. *Journal of Applied Bacteriology* **49**, 305–314.

WONG, W. C. & PREECE, T. F. 1981 *Pseudomonas tolaasii* in mushroom crops: numbers of the bacterium and measurement of symptom development on mushrooms grown in various environments after artificial inoculation. *Journal of Applied Bacteriology* submitted for publication.

Wong, W. C., Nash, T. H. & Preece, T. F. 1974 A field survey of watermark disease of the cricket bat willow in Essex and observations on some of the probable sources of the disease. *Plant Pathology* **23**, 25–29.

Zucker, M. & Hankin, L. 1970 Regulation of pectate lyase synthesis in *Pseudomonas fluorescens* and *Erwinia carotovora*. *Journal of Bacteriology* **104**, 13–18.

Interaction of Wall-free Prokaryotes with Plants

M. J. DANIELS, D. B. ARCHER AND W. P. C. STEMMER

John Innes Institute, Colney Lane, Norwich, UK

Contents

1. Introduction

THE HISTORY of plant-associated wall-free prokaryotes, generally referred to as mycoplasmas without prejudice to their taxonomic status, is short. In 1967 Doi *et al.* showed that plants infected with certain 'yellows' diseases, previously assumed to be caused by viruses, contained in their phloem bodies resembling mycoplasmas, and Ishiie *et al.* (1967) showed that the diseases responded to tetracycline therapy, thereby implicating prokaryotes as the disease agents.

From this point interest in plant mycoplasmas grew rapidly. Many papers described the presence of 'mycoplasma-like organisms' in diseased plants and in insect vectors, but attempts to culture the presumptive pathogens *in vitro* were usually unsuccessful. To date, Koch's postulates for proving the pathogenicity of cultured mycoplasmas for plants have been fulfilled for only two diseases: citrus stubborn and corn stunt. In both cases the pathogens have turned out to be members of a new group of mycoplasmas called spiroplasmas. Many other spiroplasma isolates are now known from plants and arthropods, but so far there is no reason to assume that all yellows diseases are caused by spiroplasmas.

Early work on plant mycoplasmas has been reviewed many times, most recently by Saglio & Whitcomb (1979). In this paper we shall not review the bulk of the literature but rather concentrate on aspects which have so far remained largely unrecorded.

2. The Diversity of Plant Mycoplasma Habitats

A. Phloem

Electron micrographs of yellows-diseased plants show mycoplasmas confined to phloem tissue, usually in the sieve tubes but also occasionally in companion cells. There have been some claims that other tissue can become colonized but, as McCoy (1979) pointed out, the evidence presented was not compelling and the conclusions stemmed probably from misidentification of plant tissues, or from confusion of plant organelles with mycoplasmas. Mycoplasmas are frequently found in highest concentrations in phloem of growing points. Laflèche & Bové (1970) carried out an electron microscopic study of stubborn-diseased citrus plants and were only able to detect spiroplasmas in young shoots. Using the enzyme-linked immunosorbent assay (ELISA) serological technique to measure levels of spiroplasma antigen throughout diseased plants, much higher levels were found in tips of shoots (Saillard *et al.* 1978). Immature sieve tubes in growing tips may still contain organelles (nucleus, mitochondria, etc.), and the presence of mycoplasmas in these cells may mislead the investigator into thinking that mycoplasmas have colonized parenchyma tissue. Figure 1 shows a section through an immature sieve tube of a mycoplasma-infected Siberian wallflower plant in which the cytoplasm is packed with mycoplasmas, together with plant cell organelles.

The primary reason for mycoplasmas being located in sieve tubes is that they are deposited there by feeding insect vectors (leafhoppers). Whether they are subsequently restricted to the phloem by mechanical barriers, or because other tissues offer an unfavourable environment is not known, but the failure of attempts to inoculate mycoplasmas into plants by mechanical means suggests that functioning undamaged phloem is required for their establishment.

B. Nectar

Following the discovery of a lethal spiroplasmosis of honey bees (Clark, 1977), a search was made for spiroplasmas in the nectar of flowering trees on which the bees were at the time feeding. A number of strains have been isolated from various flowers, probably belonging to several groups, and in addition non-helical mycoplasmas have been recovered (Davis *et al* 1977;

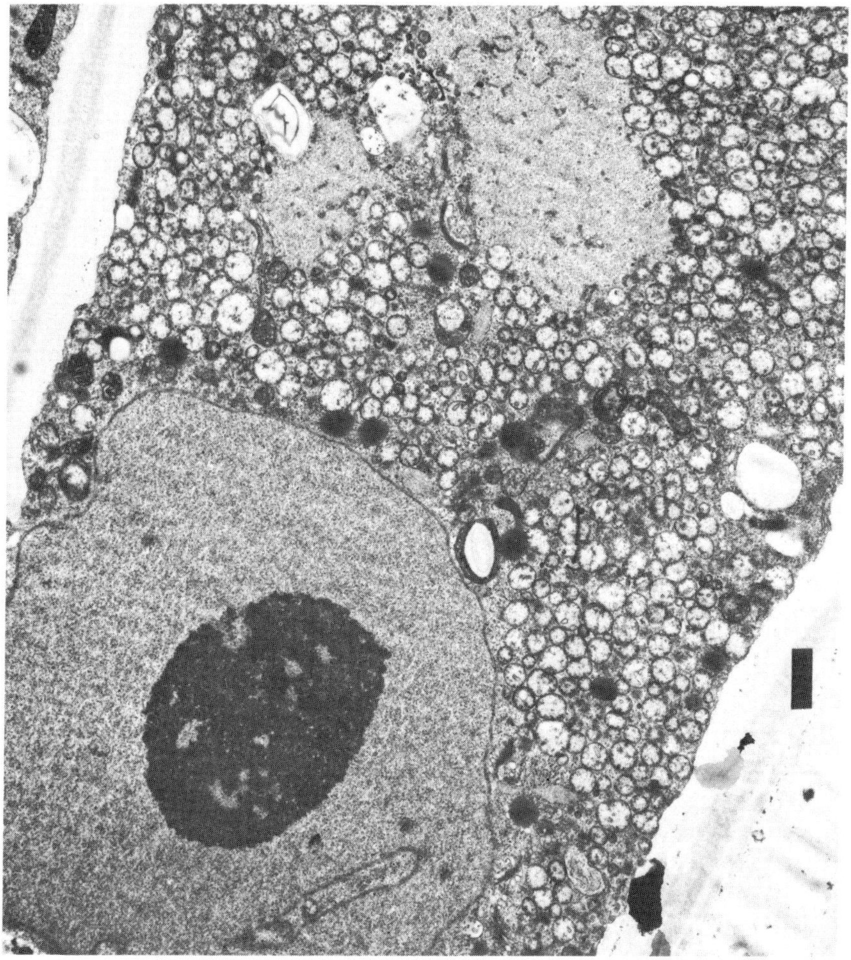

Fig. 1. Mycoplasma-infected plant (Siberian wallflower). Electron micrograph of an immature phloem cell showing cytoplasm packed with mycoplasmas. Bar = 1 μm.

Davis 1978). Plants from which these organisms are isolated show no signs of disease. Although nectar derives from phloem sap there are no direct channels from the nectary into the sieve tube so that it is unlikely that mycoplasmas could pass freely between the two tissues. Although some of the nectar spiroplasmas are pathogenic to insects, there are no reports yet of pathogenicity to plants.

C. Necrotic tissue

Lethal yellowing of palms is a disease of presumptive mycoplasma aetiology, on the basis of electron microscope studies and tetracycline therapy. One of the symptoms of the disease is necrosis and rotting of the growing crown of the palm. In the course of attempts to culture the aetiological agent of the disease, Eden-Green & Tully (1979) isolated many acholeplasma strains, notably *Acholeplasma axanthum*, from necrotic tissue. There is no evidence that the acholeplasmas are plant pathogens. Rather, it is likely that they are carried by insects feeding on the tissue. These findings have re-awakened interest in acholeplasma ecology; if they are principally insect mycoplasmas this may explain their widespread distribution in nature (Tully 1979).

D. Insects

Plant-pathogenic mycoplasmas are transmitted persistently by vector insects, which are usually leafhoppers or occasionally planthoppers. Insects acquire the organisms by feeding on phloem contents of diseased plants (or in the laboratory by feeding on suspensions of mycoplasmas, or following micro-injection of suspensions into the body cavity). A latent period follows during which the mycoplasmas cross the insect gut wall, circulate in the haemolymph, and invade and grow in other organs, reaching particularly high levels in salivary glands, whence they are discharged with saliva when the insect feeds. Growth of mycoplasmas in insects can be measured directly in the case of cultivable spiroplasmas or indirectly by autoradiographic procedures (Maillet & Gouranton 1971). In salivary gland cells mycoplasmas are enclosed in membrane-bound vacuoles and in this situation spiroplasmas appear to lose their helical shape (Townsend *et al.* 1977*a*).

3. Comparative Properties of Plant Mycoplasmas

As noted above the only plant-pathogenic mycoplasmas to have been cultured are the spiroplasmas. Indeed the discovery of this group of wall-free prokaryotes came about when the agents of citrus stubborn and corn stunt disease were first studied. Spiroplasmas are motile, helical filaments, about 0·15 μm wide and up to 5 μm long. They require sterol for growth and have a genome size of about 10^9 daltons (comparable with acholeplasmas). Their biology has been discussed by Bové & Saillard (1979). Recent taxonomic studies suggest that the corn stunt and honey bee spiroplasmas may be subspecies of *Spiroplasma citri*, the agent of citrus stubborn disease, and the first spiroplasma to be fully characterized. However, spiroplasmas from nectar fall into three groups. One is similar to the bee spiroplasma (and hence related to *S. citri*), but the others are probably two separate species (Junca *et al.*, 1980).

In the majority of yellows-diseased plants, electron microscopy shows that the mycoplasmas are not helical filaments but apparently ovoid, pleomorphic bodies similar to animal mycoplasmas. In some cases serial sectioning techniques point to a more complex morphology of branching filaments (Waters & Hunt, 1980). However, spiroplasmas may under certain conditions lose their helical shape, e.g. in old batch cultures or in insect tissues, so it may be unwise to conclude that the majority of plant mycoplasmas are not spiroplasmas. Because the organisms have not been cultured, and only one report has appeared describing their purification from diseased plants, nothing is known of their properties.

Non-helical mycoplasmas from nectar have been reported to include both sterol-requiring and non-sterol-requiring strains, but nothing has yet been published about their properties.

Acholeplasmas from necrotic palm tissue proved to be similar to strains isolated from animal sources for many years. The wide distribution in nature of some acholeplasmas had always been puzzling and the recent work suggests that they may be spread by insects which visit not only plants but also animals. The spread may be passive (non-persistent) but certain of the palm-derived strains are capable of multiplication in insect tissues (Eden-Green *et al.* 1980) so persistent carriage is also a possibility.

4. Growth of Mycoplasmas in Plants

The sieve-tube is a protected environment with all the nutrients required for the growth and reproduction of plant mycoplasmas. Spiroplasmas are, like all mycoplasmas, nutritionally exacting when grown *in vitro* but the large majority of plant mycoplasmas remain uncultivated. Formulation of media containing phloem sap, or media with compositions approximating to that of phloem sap are obvious approaches to successful cultivation of plant mycoplasmas. In general, however, only small amounts of phloem sap can be obtained, limiting its use in media, although sufficient has been obtained for chemical analyses, and so comparison of the chemical composition of phloem saps with the known nutritional requirements of mycoplasmas can be made.

Phloem sap has been collected in largest quantities as plant exudates after cutting, although it may become contaminated with constituents from other tissues (Eschrich & Heyser 1975). Van Die & Tammes (1975) have argued, however, that such plant exudates derive almost totally from the phloem and coconut 'toddy' can be collected in large amounts and is suited for use in growth media for cultivation of mycoplasmas. Small quantities of phloem sap have been obtained by the aphid stylet technique (Kennedy & Mittler 1953; Downing & Unwin 1977) and in some cases aphid honeydew has been used for nutrient analysis (Mittler 1958). Knowledge of phloem sap composition has been reviewed (Eschrich 1970; Eschrich & Heyser 1975; Ziegler 1975) and aspects relevant to growth of mycoplasmas are discussed below.

No plant mycoplasma has been grown in a chemically defined medium and so, unlike *Mycoplasma mycoides* and *Acholeplasma laidlawii*, their precise nutritional requirements are unknown although some information is available (Bové & Saillard 1979). Inorganic requirements of mycoplasmas are not well documented but most phloem saps contain levels of K^+, Mg^{2+} and PO_4^{3-} which would support good growth of mycoplasmas (Rodwell & Mitchell 1979). Organic acids present in phloem saps may not be essential nutrients for all plant mycoplasmas and are not required by *S. citri* but supplementation of media with α-ketoglutaric acid enhances the growth of the corn stunt spiroplasma (Jones *et al.* 1977). The bulk of phloem sap dry matter is carbohydrate with sucrose the major component, a sugar not known to be metabolized by mycoplasmas but a normal constituent of media used in plant mycoplasma cultivation in order to create a high osmotic pressure. Other sugars found in phloem saps include sucrose glycosides, sugar alcohols and sugar phosphate depending on the plant species (Ziegler 1975) but not normally reducing sugars although glucose, among others, has been detected (Eschrich & Heyser 1975; Van Die & Tammes 1975; W. P. C. Stemmer, unpublished observations, see below). There are many essential amino acids required for growth of mycoplasmas (Rodwell & Mitchell 1979) and although there is no information on amino acid requirements of plant mycoplasmas, the levels and range of amino acids present in most phloem saps would support good growth of mycoplasmas. Sap proteins could also provide essential amino acids for nutrition of plant mycoplasmas. Many proteins have been detected in phloem exudates, the major protein in many cases being 'P-protein'.

The preceding discussion indicates that phloem sap is indeed a suitable environment for growth of mycoplasmas, but the source of the supply of essential vitamins, lipids and nucleic acid bases is harder to understand. There is little information on vitamin requirements of plant mycoplasmas, but some water-soluble vitamins have been detected in phloem exudates (Ziegler & Ziegler 1962). Spiroplasmas, in common with *Mycoplasma* spp., require fatty acids and sterols for growth in synthetic media, but there is no evidence for substantial quantities of these lipophilic compounds in phloem sap although free fatty acids and sterols have been detected in some saps and in aphid honeydew (Strong 1965; Ziegler 1975). There is no support for the hypothesis of growth of plant mycoplasmas in intimate association with plant membranes which could conceivably provide the means for sterol acquisition. Plant mycoplasmas might grow *in vivo* with very little or no membrane sterol or, as seems more likely, near the sites of phloem differentiation where, due to structural re-organization of the plant cell, more lipids might be available. Phloem differentiation may also provide essential nucleotide bases because, apart from adenine derivatives, only small quantities of uracil and, to an even lesser extent, cytidine derivatives have been detected in sap (Ziegler 1975). Support for the preponderance of plant mycoplasmas in young leaves comes from electron microscopy (Laflèche & Bové 1970) and ELISA (Saillard *et al.* 1978). We estimate from electron micrographs that infected sieve-tube elements can contain up to 10^{13} mycoplasmas/ml. Growth and reproduction to such a high density requires continual supply of nutrients and removal of toxic products (e.g. acid) so that the phloem of growing points may be likened to a chemostat for culture of mycoplasmas.

5. Spread of Mycoplasmas in Plants

Natural transmission of mycoplasmas to plants is by insects or the parasitic dodder (McCoy 1979). Experimental transmission by grafting of dicotyledonous plants is normally successful but mechanical inoculation has not yet been achieved. Spread of mycoplasmas within the plant from its point of inoculation presumably occurs via the phloem, and mycoplasmas can be detected throughout the plant by serological techniques (when there is an antiserum), by culture or by electron microscopy. For uncultivated plant mycoplasmas electron microscopy is the only means of detection unless antisera specific for uncultivated mycoplasmas can be raised by using diseased plant extracts as the initial antigen.

Plant mycoplasmas are able to pass through sieve pores quite freely but evidence for mycoplasmas in phloem parenchyma or companion cells has been discounted (McCoy 1979). The forces responsible for mycoplasma spread within phloem are of great interest. For instance, do they spread via phloem translocation mechanisms or, in the case of spiroplasmas, are motility and chemotaxis contributory factors? What factors are known to affect the spread of mycoplasmas within a plant?

Radioactive tracer studies have shown that translocation rates of assimilated compounds generally lie between 2 and 250 cm/h (Wardlaw

1974) but translocation of a mycoplasma need not be at all related. The spread of viruses within plants may be a better analogy and virus translocation can occur at rates comparable with those for assimilates (Schneider 1965), but in the absence of any data concerning translocation of mycoplasmas analogies are of little value. Mycoplasmas are large compared with viruses and the organism found in sieve tubes of coconut palms with lethal yellowing disease varies in form and may be up to 16 μm long (Waters & Hunt 1980).

Spiroplasmas are motile organisms capable of speeds up to 18 cm/h (Daniels *et al.* 1980). Speeds of movement are dependent upon a number of factors including pH and viscosity of the medium. Most phloem saps are slightly alkaline (Ziegler 1975) which is suited to growth of spiroplasmas in synthetic media, and maximum speed of motility occurs at neutral pH (Daniels *et al.* 1980). The viscosity of phloem exudates is determined mainly by the sugar concentration and values between 1 and 2 cP are found (Ziegler 1975) so we estimate that the speed of spiroplasmas in phloem sap would be at least 3·6 cm/h, i.e. comparable with the translocation rates of slow moving assimilates but, if unidirectional and without hindrance, quite sufficient for rapid dissemination of spiroplasmas within plants. Vectorial movement of spiroplasmas occurs in response to chemical stimuli (Daniels & Longland 1980) so that dissemination of spiroplasmas within a plant from a point inoculum, if occurring as a result of motility, may be in response to chemotaxis. However, motility and chemotaxis cannot explain the dissemination within plants of most mycoplasmas, which do not have a helical morphology. Helicity and motility of spiroplasmas are related properties and a non-helical variant of *S. citri* when injected into bean plants by leafhoppers induced the same progression of symptoms found with the normal helical strain (Townsend *et al.* 1977*b*). Comparison of the distributions and rates of dissemination of helical and non-helical strains of *S. citri* has not yet been performed so the possible involvement of motility and chemotaxis of spiroplasmas within plants can only be speculated upon. Translocation of non-motile plant mycoplasmas does occur but chemotaxis and motility of spiroplasmas may enable them to locate preferentially at sites with optimum nutrient contents. Preferred locations of non-motile mycoplasmas would result from increased growth at sites with optimum nutrient levels.

Cultured organisms may be used to raise specific antisera to enable detection of the organisms in plant extracts. ELISA is a sensitive technique which has been used successfully to detect spiroplasmas in plants (Clark *et al.* 1978). We have used ELISA to monitor the distribution of *S. citri* in periwinkle (*Vinca rosea*) during the course of infection after experimental transmission by grafting. Shoot tips from shoots with and without grafts, and from root tips, were analysed by ELISA for the presence of *S. citri* antigens. *Spiroplasma citri* antigens appeared first in the grafted shoots after a few days and equivalent cell numbers in the shoots increased exponentially to a maximum level of about 10^9 spiroplasmas/g of tissue (wet weight). Extrapolation back from the detection limit suggested that *S. citri* antigens would have appeared in the shoots almost immediately after grafting, consistent with the expected translocation rates. In shoot tips without a graft, *S. citri* antigens were detected one to two weeks after grafting and some days later were detected in root tips. Equivalent cell numbers of *S. citri* in shoot and root tips again reached maximum values of about 10^9/g of tissue (wet weight). The first signs of symptom expression (Markham & Townsend 1974) appeared at the onset of detection of *S. citri* antigens although full symptom expression occurred somewhat later.

6. Parameters Affecting Spread and Growth

Many factors could conceivably affect spread and growth of mycoplasmas within plants. The scanty data on this has relied almost exclusively on symptom expression rather than detection of organisms, but some workers have used electron microscopy for detection. For those cultivable plant mycoplasmas, a means of detection such as ELISA is amenable for use in experiments designed to assess the effects of parameters such as temperature, light and plant nutrition on the spread of organisms within plants, but such work has yet to be performed. The nature of the host plant (Calavan & Oldfield 1979) can affect the symptomatology, but there is no information on the involvement of plant defence mechanisms (e.g. phytoalexins). Resistance of plants to mycoplasma-induced diseases has been reported but little is known of the mechanisms involved (Daniels 1979*a*).

Spiroplasmas have a narrow temperature range for growth and this is reflected also in their temperature sensitivity for proliferation within plants (McCoy 1979). Adverse conditions for the host plant, such as cold or poor light, affect the spread and growth of encumbent mycoplasmas. For instance, during winter mycoplasmas associated with deciduous woody plants are located preferentially in roots (Braun & Sinclair 1976) and dark treatment of aster-yellows-infected plants resulted in degeneration of the mycoplasmas (Chen & Hiruki 1977). Both winter-induced dormancy and dark eliminate photosynthetic activity so that the sieve tube environment would become depleted of its nutrient supply and unfavourable for growth of mycoplasmas.

The interactions of mycoplasmas with host plants are complex and not well understood at the molecular level. Successful cultivation of more plant pathogenic mycoplasmas will mark a major advance in this field of study but understanding the interaction of the plant with the mycoplasma is the real challenge.

7. Biochemistry of Symptom Production in Diseased Plants

The limited number of studies in this area have mostly been undertaken with spiroplasmas because of the obvious advantages of being able to compare the behaviour of the pathogen *in vivo* and *in vitro* and to dissociate the activities of the members of the host–parasite complex.

The symptoms produced by spiroplasmas on plants of many families have been recently described by Calavan & Oldfield (1979) and will not be described here in detail. The important point is that *S. citri* produces wilting, followed by death. Roots of plants at this terminal stage of disease have degenerated (Fig. 2), so that wilting probably results from impaired water uptake, a supposition which receives support from the ability of wilted shoots to regain turgor when cut from the plant and placed in water.

Fig. 2. Roots of periwinkle plants grown in hydroponic culture. Right, healthy plant; left, plants of same age 26 days after graft inoculation with *S. citri*, showing degeneration and necrosis. Bar = 5 cm.

The simplest explanation of this behaviour is that the spiroplasmas produce a diffusible toxin, and root cells are the most sensitive target for the toxin. Evidence has been obtained for the existence of such a toxin (Daniels 1979*a*), which is a low molecular weight, polar, acidic substance, as yet uncharacterized because of its chemical instability.

The corn stunt spiroplasma does not induce wilting and does not appear to produce the toxin in culture. Instead it produces symptoms more typical of the majority of yellows diseases, yellow foliage, stunting and floral abnormalities. Yellowing of leaves, although one of the most striking symptoms, has been scarcely studied at all. Loss of chlorophyll could result from a reduction of availability of nutrients (many deficiency diseases also cause chlorosis), perhaps because spiroplasmas sequester certain substances or because they interfere with phloem transport, or alternatively they may secrete substances which promote chlorophyll breakdown. A search for substances in spiroplasma cultures which cause chlorophyll breakdown in leaf discs has so far proved fruitless (Daniels & Longland, unpublished). Induction of chlorosis in host plants may have some selective value for spiroplasmas because insect vectors are attracted to yellow, and so dispersal is promoted.

Stunting of infected plants takes the form of shorter internodes and smaller organs (e.g. leaves and flowers). Maramorosch (1957) noted that gibberellin spraying reversed the stunting of maize plants caused by corn stunt, and Dabek (1974) found that gibberellin treatment suppressed foliar symptoms of lethal yellowing of palms. We found that gibberellic acid caused lengthening of internodes when applied to corn stunt-infected periwinkle plants, but other disease symptoms were unaffected (Daniels 1979*b*). We considered that the results did not justify the conclusion that spiroplasmas interfere with gibberellin metabolism, and, indeed, attempts to obtain direct evidence for this, e.g. inactivation of gibberellins, inhibition of action or of biosynthesis, gave inconclusive results. Recently, Shepardson & McCrum (1979) reported that gibberellic acid intensifies aster yellows symptoms in infected periwinkles, and moreover produces similar changes in healthy plants. It has been reported that kinetin can reverse the symptoms of potato witches' broom disease (Plavšić *et al.* 1978), but this treatment had no effect on corn stunt-diseased periwinkles (Daniels & Longland, unpublished).

Changes in chemical composition of plants as symptoms develop are not well documented. Rodeia & Borges (1973) described changes in phosphorus and anthocyanins in tomatoes with 'Mal Azul' disease, and Bové *et al.* (1961) found changes in organic acid levels in stubborn diseased citrus. Since mycoplasmas are localized in sieve tubes, changes in phloem sap are likely to be more pronounced than gross changes in composition of the

TABLE 1
Composition of coconut phloem sap

	Sap from					
	Healthy palm		Diseased palm		Healthy inflorescence	
Total solids						
(mg solid/ml sap)	62	(4)	86	(24)	153	(28)
Sugar composition* (mg sugar/g solids)						
Sucrose	561	(57)	592	(58)	599	(25)
Inositol A	101	(21)	82	(14)	2·5	(0·5)
Glucose	29	(37)	23	(30)	71	(36)
Fructose	12	(15)	18	(22)	59	(24)
Inositol B	22	(10)	13	(3)	0·5	(0·0)
Mannitol	8·1	(7·1)	5·1	(8·2)	5·0	(0·9)
Oxysuccinate	5·2	(4·8)	4·1	(3·9)	—	
Deoxyglucose	1·7	(1·3)	3·0	(1·2)	0·5	(0·1)
Raffinose	2·7	(3·7)	1·5	(3·1)	6·3	(5·4)
TOTAL	744	(13)	742	(11)	756	(12)
Amino acid composition (mg amino acid/g solids)						
Aspartic acid	4·31	(3·94)	4·08	(1·64)	0·41	(0·10)
Threonine	8·59	(6·40)	14·88	(5·68)	0·33	(0·16)
Serine	3·56	(1·93)	4·85	(0·86)	0·29	(0·06)
Glutamic acid	9·34	(1·37)	11·20	(2·24)	1·10	(0·11)
Proline	5·15	(1·47)	6·58	(1·28)	—	
Glycine	0·33	(0·10)	0·22	(0·06)	0·05	(0·01)
Alanine	4·50	(1·70)	2·56	(0·83)	0·07	(0·01)
Valine	0·96	(0·53)	1·98	(0·74)	0·02	(0·03)
Methionine	0·13	(0·07)	0·23	(0·09)	—	
Isoleucine	0·23	(0·07)	1·32	(0·72)	0·01	(0·01)
Leucine	0·25	(0·14)	0·49	(0·15)	—	
Tyrosine	0·23	(0·11)	0·30	(0·09)	—	
Phenylalanine	1·23	(0·37)	1·26	(0·50)	0·02	(0·02)
Lysine	0·23	(0·07)	0·48	(0·22)	0·02	(0·03)
Histidine	0·33	(0·23)	1·19	(0·61)	—	
Arginine	0·25	(0·10)	0·69	(0·35)	0·41	(0·21)
Half cysteine	0·01	(0·02)	0·02	(0·05)	—	
Ammonia	0·69	(0·45)	0·95	(0·28)	0·31	(0·03)
TOTAL	40·32	(12.97)	53·09	(5·87)	3·05	(0·59)
Element composition (μg/g solids)						
K	45300	(6500)	35900	(8800)	4850	(30)
Na	2210	(800)	1740	(620)	2300	(60)
Mg	970	(200)	970	(200)	240	(20)
Ca	140	(80)	220	(130)	36	(6)
Zn	27	(9)	29	(5)	5·8	(1·8)
Cu	15	(7)	19	(11)	2·7	(0·3)
Fe	15	(14)	12	(9)	11	(1)

*This section includes non-sugar CHO compounds.
Figures in brackets are standard deviations.
—, not detected

whole plant, and are more likely to reflect the activities of the mycoplasmas. Phloem sap is not easy to collect from most plants, but fortunately palms yield phloem sap, called 'toddy', abundantly (see Section 4). Samples of toddy from healthy and lethal yellowing-diseased palms collected in Jamaica have been analysed for sugars, amino acids and metals. Some typical results are shown in Table 1. No large scale differences were seen, although total sugar content apparently increased as the disease advanced in severity. An unexpected finding was that toddy collected from inflorescences of healthy palms had a qualitatively different composition from trunk toddy, lacking inositol. It has always been surprising that spiroplasmas are unable to metabolize sucrose, the major phloem sugar, but the analyses show the presence of significant quantities of metabolizable reducing sugars.

Our aim is to understand in detail the biochemical events which take place when mycoplasmas infect plants to produce disease symptoms. As in all host–parasite combinations we are faced with a complex interacting system in which the metabolism of the separate components may be diverted to new pathways in response to changes produced by the other member. Thus the behaviour of the isolated members, i.e. healthy plants or cultured spiroplasmas, may not be an infallible guide. But then, for the majority of plant mycoplasma diseases, we are denied the luxury of experimentation with cultured pathogens.

8. References

BOVÉ, J. M. & SAILLARD, C. 1979 Cell biology of spiroplasmas. In *The Mycoplasmas* Vol. 3, ed. Whitcomb, R. F. & Tully, J. G. pp.83–153. New York: Academic Press.

BOVÉ, C., MOREL, G., MONIER, F. & BOVÉ, J. M. 1961 Chemical studies on stubborn-affected Marsh grapefruit and Washington navel oranges. In *Proceedings of the Second International Organisation of Citrus Virologists* ed. Price, W. C. pp.60–68. Gainesville: University of Florida Press.

BRAUN, E. J. & SINCLAIR, W. A. 1976 Histopathology of phloem necrosis in *Ulmus americana*. *Phytopathology* **66**, 598–607.

CALAVAN, E. C. & OLDFIELD, G. N. 1979 Symptomatology of spiroplasmal plant diseases. In *The Mycoplasmas* Vol. 3 ed. Whitcomb, R. F. & Tully, J. G. pp.37–64. New York: Academic Press.

CHEN, M-H. & HIRUKI, C. 1977 Effects of dark treatment on the ultrastructure of the aster yellows agent *in situ*. *Phytopathology* **67**, 321–324.

CLARK, M. F., FLEGG, C. L., BAR-JOSEPH, M. & ROTTEM, S. 1978 The detection of *Spiroplasma citri* by enzyme-linked immunosorbent assay (ELISA). *Phytopathologische Zeitschrift* **92**, 332–337.

CLARK, T. B. 1977 *Spiroplasma* sp., a new pathogen in honey bees. *Journal of Invertebrate Pathology* **29**, 112–113.

DABEK, A. J. 1974 Biochemistry of the lethal yellowing disease of coconut palms in Jamaica. *Phytopathologische Zeitschrift* **81**, 346–353.

DANIELS, M. J. 1979*a* Mechanisms of spiroplasma pathogenicity. In *The Mycoplasmas* Vol. 3, ed. Whitcomb, R. F. & Tully, J. G . pp.209–227. New York: Academic Press.

DANIELS, M. J. 1979*b* The pathogenicity of mycoplasmas for plants. *Zentralblatt für Bakteriologie, Parasitenkunde, Infektionskrankheiten und Hygiene. Abt. 1* **245**, 184–199.
DANIELS, M. J. & LONGLAND, J. M. 1980 Chemotaxis by spiroplasmas. *Society for General Microbiology Quarterly* **7**, 85.
DANIELS, M. J., LONGLAND, J. M. & GILBART J. 1980 Aspects of motility and chemotaxis in spiroplasmas. *Journal of General Microbiology* **118**, 429–436.
DAVIS, R. E. 1978 Spiroplasma associated with flowers of the tulip tree (*Liriodendron tulipifera* L.). *Canadian Journal of Microbiology* **24**, 954–959.
DAVIS, R. E., WORLEY, J. F. & BASCIANO, L. K. 1977 Association of spiroplasma and mycoplasma-like organisms with flowers of tulip tree (*Liriodendron tulipifera* L.). *Proceedings of the American Phytopathological Society* **4**, 185–186.
DOI, Y., TERANAKA, M., YORA, K. & ASUYAMA, H. 1967 Mycoplasma- or PLT-like microorganisms found in the phloem elements of plants infected with mulberry dwarf, potato witches' broom, aster yellows or paulownia witches' broom. *Annals of the Phytopathological Society of Japan* **33**, 259–266.
DOWNING, N. & UNWIN, D. M. 1977 A new method for cutting the mouth-parts of feeding aphids, and for collecting plant sap. *Physiological Entomology* **2**, 275–277.
EDEN-GREEN, S. J. & TULLY, J. G. 1979 Isolation of *Acholeplasma* spp. from coconut palms affected by lethal yellowing disease in Jamaica. *Current Microbiology* **2**, 311–316.
EDEN-GREEN, S. J., MARKHAM, P. G. & TOWNSEND, R. 1980 Acholeplasmas and lethal yellowing disease. II. Transmission experiments. *Proceedings of the Fourth Meeting of the International Council on Lethal Yellowing*. University of Florida Publication FL-80-1, pp.10–11.
ESCHRICH, W. 1970 Biochemistry and fine structure of phloem in relation to transport. *Annual Review of Plant Physiology* **21**, 193–214.
ESCHRICH, W. & HEYSER, W. 1975 Biochemistry of phloem constituents. In *Encyclopedia of Plant Physiology* Vol. 1, ed. Zimmerman, M. H. & Milburn, J. A. pp.101–136. Heidelberg: Springer-Verlag.
ISHIIE, T., DOI, Y., YORA, K. & ASUYAMA, H. 1967 Suppressive effects of antibiotics of tetracycline group on symptom development of mulberry dwarf disease. *Annals of the Phytopathological Society of Japan* **33**, 267–275.
JONES, A. L., WHITCOMB, R. F., WILLIAMSON, D. L. & COAN, M. E. 1977 Comparative growth and primary isolation of spiroplasmas in media based on insect tissue culture formulations. *Phytopathology* **67**, 738–746.
JUNCA, P., SAILLARD, C., TULLY, J., GARCIA-JURADO, O., DEGORCE-DUMAS, J-R., MOUCHES, C., VIGNAULT, J-C., VOGEL, R., MCCOY, R., WHITCOMB, R. F., WILLIAMSON, D., LATRILLE, J. & BOVÉ, J. M. 1980 Characterisation de spiroplasmes isolés d'insectes et de fleurs de France continentale, de Corse et du Maroc; Proposition pour une classification de spiroplasmes. *Comptes rendus hebdomadaires des Séances de l'Academic des Sciences, Paris, Série D* **290**, 1209–1212.
KENNEDY, J. S. & MITTLER, T. E. 1953 A method for obtaining phloem sap via the mouthparts of aphids. *Nature, London* **171**, 528.
LAFLÈCHE, D. & BOVÉ, J. M. 1970 Mycoplasmes dans les agrumes atteints de 'greening', de 'stubborn' ou de maladies similaires. *Fruits* **25**, 455–465.
MCCOY, R. E. 1979 Mycoplasmas and yellows diseases. In *The Mycoplasmas* Vol. 3, ed. Whitcomb, R. F. & Tully, J. G. pp.229–264. New York: Academic Press.
MAILLET, P. L. & GOURANTON, J. 1971 Étude du cycle biologique du mycoplasme de la phyllodie du trèfle dans l'insecte vecteur, *Euscelis lineolatus* Brullé (*Homoptera, Jassidae*). *Journal de Microscopie* **11**, 143–162.
MARAMOROSCH, K. 1957 Reversal of virus-caused stunting in plants by gibberellic acid. *Science, N.Y.* **126**, 651–652.
MARKHAM, P. G. & TOWNSEND, R. 1974 Transmission of *Spiroplasma citri* to plants. *Les Mycoplasmes/Mycoplasmas. Les Colloques de l'Institut National de la Santé et de la Recherche Médicale* **33**, 201–206.
MITTLER, T. E. 1958 Studies on the feeding and nutrition of *Tuberolachnus salignus* (Gmelin) (*Homoptera, Aphididae*) II. The nitrogen and sugar composition of ingested

phloem sap and excreted honeydew. *Journal of Experimental Biology* **35**, 74–84.

PLAVSIĆ, B., BUTUROVIĆ, D., KRIVOKAPIĆ, K. & ERIĆ, Z. 1978 Some characteristics of mycoplasma-like infection and the effects of kinetin on the MLO-infected plants. Abstracts of the Third International Congress of Plant Pathology p.81.

RODEIA, N. & BORGES, M. DE LOURDES V. 1973 Alterations in phosphorus and anthocyanins related to the presence of mycoplasmas in tomato plants. *Portugaliae Acta Biologica Serie A* **13**, 72–78.

RODWELL, A. W. & MITCHELL, A. 1979 Nutrition, growth and reproduction. In *The Mycoplasmas* Vol. 1, ed. Barile, M. F. & Razin, S. pp.103–139. New York: Academic Press.

SAGLIO, P. H. M. & WHITCOMB, R. F. 1979 Diversity of wall-less prokaryotes in plant vascular tissue, fungi and invertebrate animals. In *The Mycoplasmas* Vol. 3, ed. Whitcomb, R. F. & Tully, J. G. pp.1–36. New York: Academic Press.

SAILLARD, C., DUNEZ, J., GARCIA-JURADO, O., NHAMI, A. & BOVÉ, J. 1978 Detection de *Spiroplasma citri* dans les agrumes et les pervenches par la technique immuno-enzymatique 'ELISA'. *Comptes rendus hebdomadaires des Séances de l'Academie des Sciences, Paris, Série D* **286**, 1245–1248.

SCHNEIDER, I. R. 1965 Introduction, translocation, and distribution of viruses in plants. *Advances in Virus Research* **11**, 163–221.

SHEPARDSON, S. & MCCRUM, R. C. 1979 Effect of gibberellic acid on mycoplasma-like organism-infected and healthy periwinkle. *Plant Disease Reporter* **63**, 865–869.

STRONG, F. E. 1965 Detection of lipids in the honeydew of an aphid. *Nature, London* **205**, 1242.

TOWNSEND, R., MARKHAM, P. G. & PLASKITT, K. A. 1977*a* Multiplication and morphology of *Spiroplasma citri* in the leafhopper *Euscelis plebejus*. *Annals of Applied Biology* **87**, 307–313.

TOWNSEND, R., MARKHAM, P. G., PLASKITT, K. A. & DANIELS, M. J. 1977*b* Isolation and characterization of a non-helical strain of *Spiroplasma citri*. *Journal of General Microbiology* **100**, 15–21.

TULLY, J. G. 1979 Special features of the acholeplasmas. In *The Mycoplasmas* Vol. 1, ed. Barile, M. F. & Razin, S. pp.431–449. New York: Academic Press.

VAN DIE, J. & TAMMES, P. M. L. 1975 Phloem exudation from monocotyledonous axes. In *Encyclopedia of Plant Physiology* Vol. 1, ed. Zimmerman, N. H. & Milburn, J. A. pp.196–222. Heidelberg: Springer-Verlag.

WARDLAW, I. F. 1974 Phloem transport: physical, chemical or impossible. *Annual Review of Plant Physiology* **25**, 515–539.

WATERS, H. & HUNT, P. 1980 The *in vivo* three-dimensional form of a plant mycoplasma-like organism by the analysis of serial ultrathin sections. *Journal of General Microbiology* **116**, 111–131.

ZIEGLER, H. 1975 Nature of transported substances. In *Encyclopedia of Plant Physiology* Vol. 1, ed. Zimmerman, M. H. & Milburn, J. A. pp.59–100. Heidelberg: Springer-Verlag.

ZIEGLER, H. & ZIEGLER, I. 1962 Die wasserlöslichen Vitamine in den Siebröhrensäften einiger Bäume. *Flora* **152**, 257–278.

The Biology of the Crown Gall — A Plant Tumour Induced by *Agrobacterium tumefaciens*

A. G. HEPBURN

John Innes Institute, Colney Lane, Norwich, UK

Contents

1. The Molecular Basis of Tumorigenicity

ALTHOUGH the causative relationship between the crown gall disease of dicotyledonous plants and the Gram negative soil bacterium *Agrobacterium tumefaciens* was first demonstrated by Smith & Townsend in 1907, it is only in the past six years that the 'tumour-inducing principle' (TIP) has been identified and isolated. Zaenen *et al.* (1974) showed that the ability of any strain of *A. tumefaciens* to transform plant cells was directly correlated with the presence in that strain of a large single-copy plasmid (the Ti plasmid). The Ti group of plasmids, which can itself be subdivided into groups based on the tumorigenicity-related functions they encode, has a size range of $90 \times 10^6 - 150 \times 10^6$ daltons. It is important to note however that the presence in any particular strain of a plasmid of this size range does not necessarily imply that the strain is tumorigenic; the majority of *A. tumefaciens* and indeed *A. radiobacter* strains contain one or more plasmids (Currier & Nester, 1976; Merlo & Nester, 1977). Figure 1 shows the plasmid profile of *A. tumefaciens* strain K24 which, in addition to a Ti plasmid identical to that of strain C58, contains at least four other large plasmids of comparable size. These non-Ti plasmids confer no known phenotypes on the host strain but appear to be partially related to one another in that recombination has been observed between them (W. P. Roberts, pers. comm.).

The Ti plasmids, on the other hand, do confer observable phenotypes on their hosts (see Kado (1980) for review). It was one of these phenotypes, opine catabolism, which first suggested both the form of the TIP and the

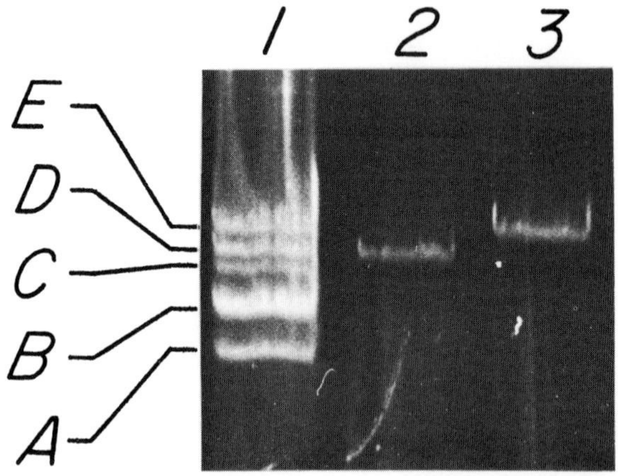

Fig. 1. 0·7% Agarose gel showing the plasmid profile of *Agrobacterium tumefaciens* strain K84 and two exconjugant derivatives. The five plasmids in strain K84 are labelled A through E (column 1). Columns 2 and 3 show exconjugant strains containing only plasmids D and E, respectively. Plasmid B is identical to the Ti plasmid pTiC58. Plasmids A and C appear to be recombinant derivatives of plasmid B. Photograph by kind permission of Dr W. P. Roberts.

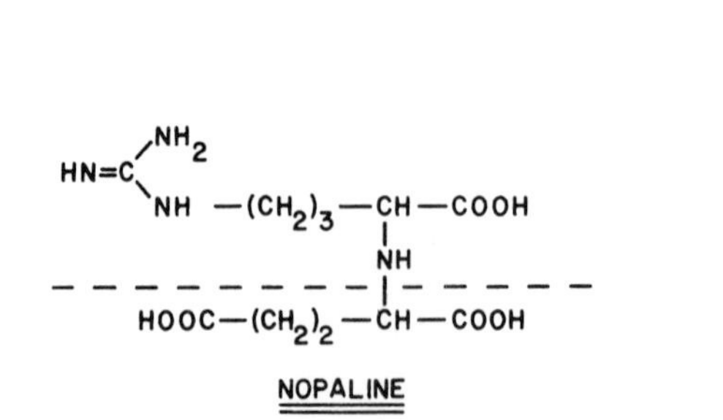

Fig. 2. The chemical structures of nopaline and octopine showing the result of the condensation of arginine with pyruvic acid (octopine) or ketoglutaric acid (nopaline).

means of plant cell transformation. The first opines to be identified were condensation products of an amino acid and a keto acid (Fig. 2) (Goldman *et al.* 1968) and were identified in *A. tumefaciens*-induced plant tumours. It is important to note that such tumours can be grown completely free of bacteria or bacteroid-like derivatives and yet these tumours synthesize opines although no such capability had ever been demonstrated in normal healthy plants. Only two enzymes are involved, one for each of the two groups of opines first identified. Octopine synthase (Hack & Kemp 1980) is a single protein of molecular weight 38,000–39,000 and is capable of condensing pyruvate with arginine, histidine, lysine or ornithine, with comparable kinetics using either NADH or NADPH as a source of reducing power. Nopaline synthase (Kemp *et al.* 1979) on the other hand comprises four apparently identical subunits each of *ca.* 40,000 mol. wt giving an estimated total molecular weight of 158,000 and condenses α-keto glutarate only with arginine or ornithine, again using either NADH or NADPH as a source of reducing power. The amino acid compositions of the two enzymes are quite similar indicating a possible close relationship between them (Hack & Kemp 1980). It was then demonstrated that a particular strain of *A. tumefaciens*, which induced tumours synthesizing, say, octopine, was capable of utilizing octopine as a sole source of carbon and nitrogen (Montoya *et al.* 1977). Once the Ti plasmids had been identified, it was soon shown that tumorigenicity, opine catabolism and opine synthesis were directly related to one another and to particular Ti plasmids (Bomhoff *et al.* 1976). The simplest explanation of this relationship was that, during transformation, specific genes were transferred from the bacterium to the plant cell. The range of 'opines' and indeed opine groups, which it is now clear should be defined operationally rather than chemically (Schell 1979), is constantly being extended. The most recently identified opine, agropine (Firmin & Fenwick 1978) (Fig. 3) appears to be the product of a condensation reaction between an amino acid and a sugar residue rather than a keto acid (Coxon *et al.* 1980).

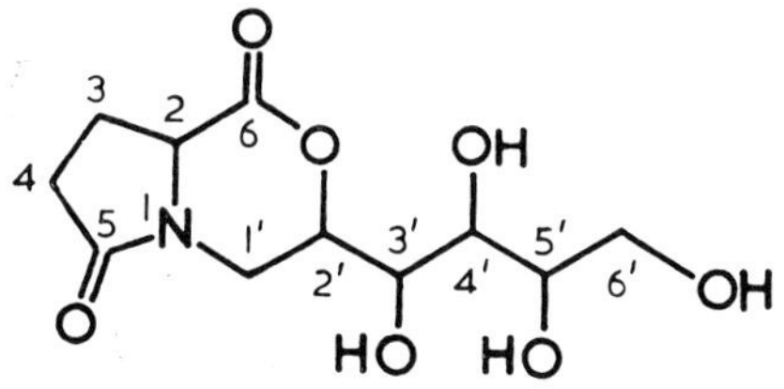

Fig. 3. The chemical structure of agropine.

As has already been mentioned, the Ti plasmids can be classified according to the opine group whose metabolism they specify. Table 1 lists the main groups identified to date.

The opines play one other major role in the life of the Ti plasmid. The Ti plasmids are conjugative plasmids (Petit *et al.* 1978) and the transfer frequency of any one plasmid is increased by several orders of magnitude in the presence of the specific opine for whose metabolism the Ti plasmid codes. The induction is Ti plasmid specific; for instance growing strain B6S3 in the presence of octopine and a suitable plasmid-less recipient strain results in the transfer of only the Ti plasmid to the recipient and at a higher frequency than occurs in the absence of octopine.

Given then that the induction of crown gall tumours results in the expression of a new genetic function whose phenotype is Ti plasmid specific rather than host plant specific, it was suggested that transformation involved the transfer of genetic information from the bacterium to the plant cell and that the source of this genetic information was the Ti plasmid. This model

TABLE 1

Ti plasmid groups defined by opine metabolism

Group*	Typical plasmids	Reference
Nopaline†	pTi C58, pTi T37	Montoya *et al.* (1977)
Octopine A-type‡	pTi B6-806, pTi A6NC (wide host range)	Montoya *et al.* (1977)
Octopine B-type	pTi Ag57, pTi Ag63 (narrow host range)	Thomashow *et al.* (1980*b*)
Agropine	pTi Bo542 (used to be the 'Null' group)	Guyon *et al.* (1980)

*With the exception of agropine for which only the anabolic and catabolic functions have been shown, the opines used to define each group are those for which the general prokaryotic catabolic, prokaryotic conjugative and eukaryotic anabolic functions have been demonstrated.
†Several minor nopaline groups have been identified: A, where the bacterium can also utilize octopine probably resulting from constitutive expression of the nopaline operon (Klapwijk *et al.* 1977); B, where no nopaline is synthesized in the tumour probably resulting from a mutation in the nopaline synthase gene exemplified by pTi AT181 (Montoya *et al.* 1977).
‡Many of the plasmids included in this group also code for agropine metabolism (Firmin & Fenwick 1978).

of plant cell transformation was confirmed by Chilton *et al.* (1977) who showed by DNA–DNA re-association studies of purified restriction endonuclease fragments of the octopine A-type plasmid pTi B6-806 in the presence and absence of tumour DNA, that a minimum of 5×10^6 daltons of DNA from a specific region of pTi B6-806 was present in several copies per cell in the plant tumour. These observations have since been considerably extended. The octopine A-type plasmids are now known to transfer two pieces of plasmid DNA to plant cells either together or separately (Thomashow *et al.* 1980*a*). One piece of *ca.* 8×10^6 daltons is always found in tumour lines at a frequency of about 1 copy/cell. The other piece of about 5×10^6 daltons can either be partially or completely absent or present at a frequency of up to 30 copies/cell. The two pieces are colinear on the restriction endonuclease map of pTi B6-806 but are usually found integrated as separate units. The nopaline-type plasmids as exemplified by pTiC58 or pTiT37 on the other hand, transfer a somewhat larger piece of plasmid DNA of *ca.* 16×10^6 daltons as a single piece which is maintained at a frequency of 1 – 20 copies/tumour cell (Holsters *et al.* 1980).

The evidence accumulated to date suggests that the plasmid DNA which is transferred during the infection process (T-DNA) is integrated into the chromosomal DNA of the transformed plant cells (Thomashow *et al.* 1980*a*) and is extensively transcribed into poly-A-containing mRNAs (Gurley *et al.* 1979; Yang *et al.* 1980*a*; McPherson *et al.* 1980). That part of this transcription is responsible for the expression of the opine synthases is suggested by Ti plasmid mutagenesis experiments which locate these genes within the T-DNA (Holsters *et al.* 1980; Koekman *et al.* 1979). Indeed McPherson *et al.* (1980) have shown that those RNA species isolated from octopine A-type plasmid induced tumours which are homologous to the octopine A-type plasmid T-DNA code, in a cell-free system, for the synthesis of three proteins of approximately 30,000, 16,500 and 15,000 daltons when the source tumours have been shown to synthesize octopine, and only one (the smallest of the three) when the source tumours lack octopine synthetic activity. It is possible that the largest protein could be an unmodified (e.g. unglycosylated) form of octopine synthase. The smallest protein, which was observed in all tumour lines examined, is translated from RNA homologous to T-DNA sequences which have been found in all Ti plasmids examined to date and therefore may play a fundamental role in the maintenance of the tumour phenotype. The possible role of the third protein is unknown but it is tempting to speculate that it may be responsible for agropine synthesis, the other opine metabolic activity encoded by octopine A-group plasmids (Firmin & Fenwick 1978).

It appears then that the transformation of dicotyledonous plant cells by *A. tumefaciens* involves plasmid-encoded functions and results in the stable integration into plant chromosomal DNA and subsequent maintenance of a small piece of plasmid DNA. It should be noted, however, that several other regions of the Ti plasmid which together with the T-DNA comprise a maximum of 45×10^6 daltons of DNA (Koekman *et al.* 1979; Holsters *et al.* 1980; Kado 1980) are involved in the total process of tumorigenicity.

2. The Infection Process

It has been demonstrated repeatedly that the infection of plants by *A. tumefaciens* has an absolute requirement for damaged or wounded plant cells (Lippincott & Lippincott 1969) but this is rather an over-simplification of the requirements for infection. It is clear from electron microscopic evidence that the primary event is the binding of the bacterium to a component of the plant cell wall (Smith & Hindley 1978). The bacterial

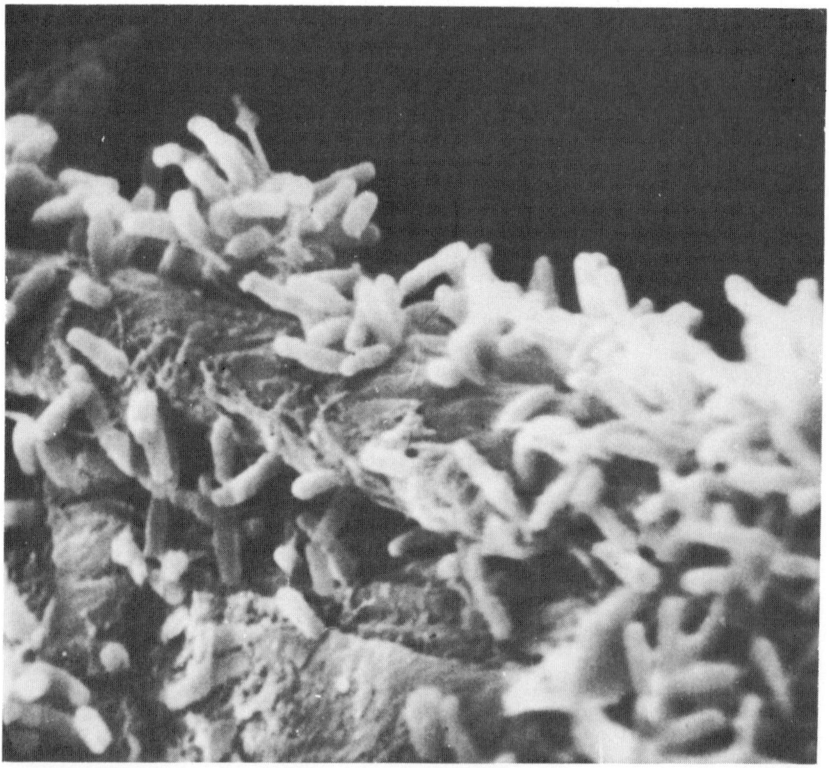

Fig. 4. Scanning electronmicrograph of *Agrobacterium tumefaciens* strain C58 attached to tobacco suspension culture cells × 8264. (Photograph taken from Smith & Hindley, 1978.)

component of this binding is under Ti plasmid control in that strain C58 cured of its Ti plasmid is unable to bind. The eukaryotic and prokaryotic components which contribute to this interaction have been at least partly elucidated by Lippincott *et al.* (1977) and Lippincott & Lippincott (1978).

The bacterial component appears to be a lipopolysaccharide constituent of the bacterial cell wall which binds to a specific component located in the plant cell wall. Evidence accumulated to date from several laboratories would indicate that this plant component is only exposed in young cell walls. Experiments with plant protoplasts which are regenerating cell walls, and tobacco cell suspension cultures, suggest that the plant cell 'acceptor' may be only available for binding for a short period during the growth of the cell wall. This may explain the need for wounding when whole plants are infected in that only when the cell walls are damaged is the plant cell binding component exposed. The binding is quite specific—the bacterial rods bind at one end and are rarely found attached singly (Fig. 4) but whether this results from a co-operative aspect of the binding or whether the clumping merely reflects a localization of the plant cell component of the interaction is not clear.

The major method of biological control of *A. tumefaciens* either directly or indirectly interferes with this binding. *Agrobacterium radiobacter* strain K84 produces a highly specific nucleoside bacteriocin, agrocin 84 (Fig. 5, Roberts *et al.* 1977) which can prevent infections produced by many

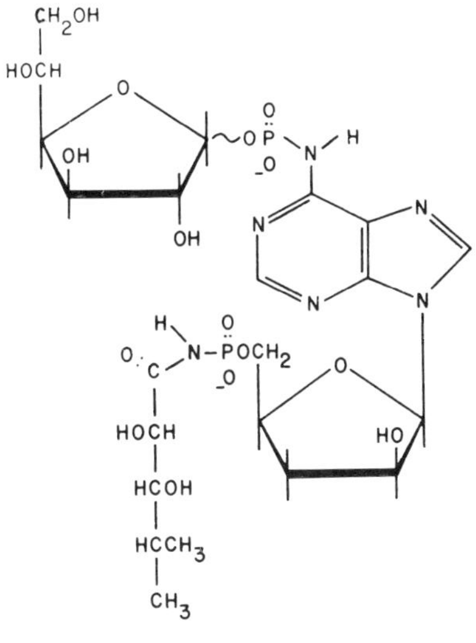

Fig. 5. The chemical structure of agrocin 84.

A. tumefaciens strains containing nopaline group Ti plasmids but none containing octopine group Ti plasmids. The agrocin 84 production gene or genes are carried by a small plasmid of 30 × 10^6 daltons (Ellis *et al.* 1979) and the sensitivity gene in pTi C58 has been mapped (Holsters *et al.* 1980). It is clear, however, that agrocin 84 production alone is only one factor which makes strain K84 a suitable organism for biological control.

To be effective in the field, an agrocin 84 producer must, at least until it is established, be able to compete with the tumorigenic strains that it is supposed to control. It is also worth noting that this type of biological control can, partly because the bacteriocin and the sensitivity are both plasmid encoded, lead to the generation of agrocin 84 resistant tumorigenic strains and indeed agrocin 84-*producing* tumorigenic strains. As a partial elucidation of the action of agrocin 84, Smith & Hindley (1978) showed that treatment with agrocin 84 prevents binding of the susceptible bacteria to the plant cell wall but it is not yet clear whether this is a primary effect of the bacteriocin or a secondary effect following on from the effect of the bacteriocin on cell growth.

The host range of *A. tumefaciens* is very extensive (De Cleene & De Ley 1976). Lower plants are not infected but the gymnosperms and the angiosperms appear to be extremely susceptible although in general monocotyledonous plants are resistant. Host range surveys have to date used only gall production as an assay and it is therefore not known whether the resistance of monocotyledonous plants, for instance, is a result of *A. tumefaciens* being unable to bind to the monocotyledonous plant cell walls, or a failure to integrate the T-DNA into the monocotyledonous plant genome. Binding assays conducted on resistant and susceptible plants could provide considerable information on this first stage of the infection process.

The next stages of the infection process are as yet unknown but it is clear that the Ti plasmid, presumably intact, enters the plant cell nucleus where it can remain for at least nine days. Nuti *et al.* (1980) have shown by *in situ* hybridization techniques using radioactively labelled Ti plasmid as a probe and hybridizing this to squashes of infected cells prepared at various times after infection, that the probe hybridizes extensively to the infected cell nuclei but that this hybridization is no longer visible 24 days after infection. The techniques used were not sensitive enough to detect the T-DNA presumably present in the infected cell genomes. Whatever the process whereby the plasmid is transferred from the bacterium to the plant cell nucleus, it is clear that the bacterium itself plays only a passive role, since it has been shown that petunia protoplasts can be transformed using purified Ti plasmid (Davey *et al.* 1980). It was observed that the normal tumour phenotype (ability to grow in the absence of plant hormones) was not stably expressed until some time after protoplast transformation which may indicate that the intact T-DNA is only capable of conferring the tumour phenotype on plant cells after integration.

The *in situ* hybridization evidence of Nuti *et al.* (1980) would appear to suggest that cells either become transformed with several plasmid copies or with only a few plasmid copies which then replicate within the plant cell nucleus. The protoplast system may be able to resolve the question of whether the Ti plasmid actually replicates in plant cells. An alternative approach is being tried by several groups who are currently investigating whether or not one of the non-T-DNA oncogenic functions could be an origin of replication capable of being expressed in eukaryotic cells.

3. Plant Regeneration and the Fate of the T-DNA

Transformed plant cells in general grow as rapidly dividing undifferentiated parenchymatous cells in the absence of exogenously applied hormones (Fig. 6*a*). However, the same growth characteristics are shown by habituated plant cell culture lines which have never been exposed to *A. tumefaciens*. True tumour cell lines contain copies of the T-DNA and synthesize the opines characteristic of the bacterial strain which induced the tumour. The tumour phenotype, however, is not always permanent. In the case of tumours produced by particular bacterial strain–plant interactions, teratomas are repeatably produced (Fig. 6*b*; Gresshoff *et al.* 1979). These teratomas can frequently be excised and cultured as apparently normal plants. The T-DNA is still present in the cells of these plants which still synthesize opines but the tumour phenotype is suppressed. Such teratomatic plants will produce tumours either spontaneously or if damaged. It is important to note, however, that primary tumours excised from infected plants are not necessarily homogeneous. Sacristan & Melchers (1977) claim that one tumour cell can provide sufficient hormones to maintain 10^2 - 10^3 normal cells and hence before plants arising from a tumour can be claimed as teratomas, it must be clearly demonstrated that they too either contain T-DNA or synthesize opines. This also emphasizes the necessity to use cloned tumour tissue in which all the cells contain the T-DNA for studies on the intracellular copy numbers of the T-DNA. It is in this context that the protoplast transformation system of Davey *et al.* (1980) may prove very important in that tumours can be isolated which are the product of the transformation of single protoplasts and as such already represent cloned tumours.

Although teratomatic plants can flower and set seeds, plants grown from such seeds lack the T-DNA and do not synthesize opines but it is not clear whether this loss is because meiosis actively eliminates the T-DNA or because only those cells which have already lost the T-DNA are capable of undergoing meiosis (Yang *et al.* 1980*b*).

Fig. 6. Examples of two extremes of tumour phenotype. (*a*) Undifferentiated tumour cells, × 2·4. (*b*) Teratomatic plantlets arising from tumour cells, × 2·2. Both tumours were originated on tobacco plants.

4. Genetic Engineering Prospects

The T-DNA of the Ti plasmid can carry non-Ti plasmid DNA into dicotyledonous plants where it is maintained through the somatic cell line. Natural insertions of transposons into the T-DNA of pTi C58 have been successfully integrated into tobacco cell genomes, indicating that at least $9{\cdot}5 \times 10^6$ daltons of foreign DNA can be incorporated into plant genomes using the Ti plasmid (Holsters *et al.* 1980). It appears feasible to transfer intact genes to a wide range of plants in the same way. Alternatively, it may be possible to insert the structural components of desirable genes into the 5' end of the nopaline synthase gene, to use the nopaline synthase gene controlling sequences to drive expression of the inserted gene in the same way that the *lac Z* expression vectors are used in *Escherichia coli* (Charnay *et al.* (1978). Although the vector normally results in the tumorous transformation of infected cells, the fact that viable teratomas can be produced indicates that stable, viable, transformed plants could perhaps be produced by modification of the Ti plasmid. The major problem in the productive genetic engineering of higher plants lies in the meiotic block which prevents the T-DNA being passed on to subsequent generations thus limiting the possibility of genetic engineering at present to asexually reproducing species. So little is known about the stability of the T-DNA within the plant cell that it is not possible to speculate on the reasons for this block and to formulate models for overcoming it.

5. References

BOMHOFF, G., KLAPWIJK, P. M., KESTER, H. C. M., SCHILPEROORT, R. A., HERNALSTEENS, J. P. & SCHELL, J. 1976 Octopine and nopaline synthesis and breakdown genetically controlled by a plasmid of *Agrobacterium tumefaciens*. *Molecular and General Genetics* **145**, 177–181.

CHARNAY, P., PERRICAUDET, M., GALIBERT, F. & TIOLLAIS, P. 1978 Bacteriophage lambda and plasmid vectors, allowing fusion of cloned genes in each of three translational phases. *Nucleic Acids Research* **5**, 4479–4494.

CHILTON, M-D., DURMMOND, M. H., MERLO, D. J., SCIAKY, D., MONTOYA, A. L., GORDON, M. P. & NESTER, E. W. 1977 Stable incorporation of plasmid DNA into higher plant cells: the molecular basis of crown gall tumorigenesis. *Cell* **11**, 263–271.

COXON, D. T., DAVIES, A. M. C., FENWICK, G. R., SELF, R., FIRMIN, J. L., LIPKIN, D. & JANES, N. F. 1980 Agropine, a new amino acid derivative from crown gall tumours. *Tetrahedron Letters* **21**, 495–498.

CURRIER, T. L. & NESTER, E. W. 1976 Evidence for diverse types of large plasmids in tumor-inducing strains of *Agrobacterium*. *Journal of Bacteriology* **126**, 157–165.

DAVEY, M. R., COCKING, E. C., FREEMAN, J., PEARCE, N. & TUDOR, I. 1980 Transformation of *Petunia* protoplasts by isolated *Agrobacterium* plasmids. *Plant Science Letters* **18**, 307–313.

DE CLEENE, M. & DE LEY, J. 1976 The host range of crown gall. *The Botanical Review* **42**, 389–466.

ELLIS, J. G., KERR, A., VAN MONTAGU, M. & SCHELL, J. 1979 *Agrobacterium:* genetic studies on agrocin 84 production and the biological control of crown gall. *Physiological Plant Pathology* **15**, 311–319.

FIRMIN, J. L. & FENWICK, G. R. 1978 Agropine—a major new plasmid-determined metabolite in crown gall tumours. *Nature, London* **276**, 842–844.

GOLDMAN, A., TEMPÉ, J. & MOREL, G. 1968 Quelques particularités de diverses souches d'*Agrobacterium tumefaciens*. *Comptes Rendues Academie de Sciences de Paris* **162**, 630–631.

GRESSHOFF, P. M., SKOTNICKI, M. L. & ROLFE, B. G. 1979 Crown gall teratoma formation is plasmid and plant controlled. *Journal of Bacteriology* **137**, 1020–1021.

GURLEY, W. B., KEMP, J. D., ALBERT, M. J., SUTTON, D. W. & COLLINS, J. 1979 Transcription of Ti-plasmid derived sequences in three octopine-type crown gall tumor lines. *Proceedings of the National Academy of Sciences, U.S.A.* **76**, 2828–2832.

GUYON, P., CHILTON, M-D., PETIT, A. & TEMPE, J. 1980 Agropine in 'null-type' crown gall tumors: evidence for generality of the opine concept. *Proceedings of the National Academy of Sciences, U.S.A.* **77**, 2693–2697.

HACK, E. & KEMP, J. D. 1980 Purificaion and characterization of the crown gall-specific enzyme, octopine synthase. *Plant Physiology* **65**, 949–955.

HOLSTERS, M., SILVA, B., VAN VLIET, F., GENETELLO, C., DE BLOCK, M., DHASE, P., DEPICKER, A., INZÉ, D., ENGLER, G., VILLARROEL, R., VAN MONTAGU, M. & SCHELL, J. 1980 The functional organization of the nopaline *A. tumefaciens* plasmid pTi C58. *Plasmid* **3**, 212–230.

KEMP, J. D., SUTTON, D. W. & HACK, E. 1979 Purification and characterization of the crown gall specific enzyme nopaline synthase. *Biochemistry* **18**, 3755–3760.

KADO, C. I. 1980 pTi plasmid of *Agrobacterium tumefaciens* as a cloning vector in the transformation of higher plants. Recombinant DNA Technical Bulletin 2, 145–153, U.S. Department of Health and Human Services, NIH Publication No. 80–99.

KLAPWIJK, P. M., OUDSHOORN, M. & SCHILPEROORT, R. A. 1977 Inducible permease involved in the uptake of octopine, lysopine and octopinic acid by *Agrobacterium tumefaciens* strains carrying virulence-associated plasmids. *Journal of General Microbiology* **102**, 1–11.

KOEKMAN, B. T., OOMS, G., KLAPWIJK, P. M. & SCHILPEROORT, R. A. 1979 Genetic map of an octopine Ti-plasmid. *Plasmid* **2**, 347–357.

LIPPINCOTT, B. B. & LIPPINCOTT, J. A. 1969 Bacterial attachment to a specific wound site as an essential stage in tumor induction by *Agrobacterium tumefaciens*. *Journal of Bacteriology* **97**, 620–628.

LIPPINCOTT, J. A. & LIPPINCOTT, B. B. 1978 Cell walls of crown-gall tumors and embryonic plant tissues lack *Agrobacterium* adherence sites. *Science* **199**, 1075–1078.

LIPPINCOTT, B. B. & WHATLEY, M. H. & LIPPINCOTT, J. A. 1977 Tumor induction by *Agrobacterium* involves attachment of the bacterium to a site on the host plant cell wall. *Plant Physiology* **59**, 388–390.

MCPHERSON, J. C., NESTER, E. W. & GORDON, M. P. 1980 Proteins encoded by *Agrobacterium tumefaciens* Ti plasmid DNA (T-DNA) in crown gall tumors. *Proceedings of the National Academy of Sciences, U.S.A.* **77**, 2666–2670.

MERLO, D. & NESTER, E. W. 1977 Plasmids in avirulent strains of Agrobacterium. *Journal of Bacteriology* **129**, 76–80.

MONTOYA, A. L., CHILTON, M-D., GORDON, M. P., SCIAKY, D. & NESTER, E. W. 1977 Octopine and nopaline metabolism in *Agrobacterium tumefaciens* and crown gall tumor cells: role of plasmid genes. *Journal of Bacteriology* **129**, 101–107.

NUTI, M. P., LEDEBOER, A. M., DURANTE, M., NUTI-RONCHI, V. & SCHILPEROORT, R. A. 1980 Detection of Ti-plasmid sequences in infected tissues by in situ hybridization. *Plant Science Letters* **18**, 1–6.

PETIT, A., TEMPÉ, J., KERR, A., HOLSTERS, M., VAN MONTAGU, M. & SCHELL, J. 1978 Substrate induction of conjugative activity of *Agrobacterium tumefaciens* Ti plasmids. *Nature, London* **271**, 570–571.

ROBERTS, W. P., TATE, M. E. & KERR, A. 1977 Agrocin 84 is a 6-N-phosphoramidate of an adenine nucleotide analogue. *Nature, London* **265**, 379–381.

SACRISTAN, M. D. & MELCHERS, G. 1977 Regeneration of plants from 'habituated' and '*Agrobacterium*-transformed' single-cell clones of tobacco. *Molecular and General Genetics* **152**, 111–117.

SCHELL, J. 1979 Crown gall: transfer of bacterial DNA to plants via the Ti-plasmid. In *Nucleic Acids in Plants* ed. Hall, T. C. & Davies, J. W. pp.195–210. Boca Raton, Florida, USA: CRC Press.

SMITH, E. F. & TOWNSEND, C. O. 1907 A plant-tumor of bacterial origin. *Science, N.Y.* **25**, 671–673.

SMITH, V. A. & HINDLEY, J. 1978 Effect of agrocin 84 on attachment of *Agrobacterium tumefaciens* to cultured tobacco cells. *Nature, London* **276**, 498–500.

THOMASHOW, M. F., NUTTER, R., MONTOYA, A. L., GORDON, M. P. & NESTER, E. W. 1980*a* Integration and organization of Ti plasmid sequences in crown gall tumors. *Cell* **19**, 729–739.

THOMASHOW, M. F., PANAGOPOULOS, C. G., GORDON, M. P. & NESTER, E. W. 1980*b* Host range of *Agrobacterium tumefaciens* is determined by the Ti plasmid. *Nature, London* **283**, 794–796.

YANG, F., MCPHERSON, J. C., GORDON, M. P. & NESTER, E. W. 1980*a* Extensive transcription of foreign DNA in a crown gall teratoma. *Biochemical and Biophysical Research Communications* **92**, 1273–1277.

YANG, F., MONTOYA, A. L., MERLO, D. J., DRUMMOND, M. H., CHILTON, M-D., NESTER, E. W. & GORDON, M. P. 1980*b* Foreign DNA sequences in crown gall teratoma and their fate during the loss of the tumorous traits. *Molecular and General Genetics* in press.

ZAENEN, I., VAN LAREBEKE, N., TEUCHY, H., VAN MONTAGU, M. & SCHELL, J. 1974 Supercoiled circular DNA in crown gall inducing *Agrobacterium* strains. *Journal of Molecular Biology* **86**, 109–127.

Bacterial Diseases of Food Plants — An Overview

CONSTANCE M. E. GARRETT

East Malling Research Station, Maidstone, Kent, UK

Contents

1. Introduction

FOR LIFE AND health man must eat. Only 300 of the 200,000 or more plant species are widely grown as food and about 20 account for 90% of the world's food supply (Thurston 1973). Modern improvements in hygiene and medicine have aided the population explosion and traditional agricultural methods are quite inadequate to ensure a satisfactory diet for all and to combat the many factors that limit efficient food production, of which plant disease is but one.

Both numerically and economically the fungi are the most important group of plant pathogens. However, some bacteria too can be very destructive, especially those occurring in tropical and subtropical regions.

The majority of known plant pathogenic bacteria (excluding spiroplasmas and rickettsias) are Gram negative and in the genera *Pseudomonas*, *Xanthomonas*, *Erwinia* and *Agrobacterium* (for an abridged classification see Billing & Garrett 1980); there are also a few Gram positive coryneform pathogens and one actinomycete, *Streptomyces scabies*. Classification at the species level is subject to frequent change (Breed *et al.* 1957; Buchanan & Gibbons 1974), is far from satisfactory for the plant bacteriologist, and creates special problems for plant health inspectors, so pathologists prefer to retain many of the earlier specific epithets (Dye *et al.* 1975; Young *et al.* 1978). A special-purpose classification for the Pseudomonadaceae reflecting the high degree of host specificity shown by closely related organisms would be widely welcomed (Billing & Garrett 1980). Specificity is determined by factors of which we are largely ignorant, but herein lie the small, but vitally important, differences between pathogens. When our knowledge of them is more complete the validity of some nomenspecies may be re-established. In this paper the earlier nomenspecies of *Pseudomonas* and *Xanthomonas* have been retained but acceptance of proposed changes (Young *et al.* 1978) may reduce these to pathovar status within *Ps. syringae* and *X. campestris*, respectively (see note on p.132).

Three plant attributes, their stationary nature, their protective surfaces and their seasonal growth periodicity, greatly influence host–pathogen relationships. Bacterial plant pathogens are dependent on the movement of plants or plant parts for their primary dispersal while secondary spread is chiefly through environmental factors such as rain and wind and, in a few diseases, by pollinating insects; man's cultural practices also frequently aid spread (Crosse 1957). Phytopathogenic bacteria must enter plants through natural openings or artificial wounds before they can become established (see Ch. 4, this volume). Most bacterial pathogens survive the period between crops or seasons within or on the perennating organs of the plant; a few are adapted to survive in insect hosts and some in the soil (Crosse 1968; Schuster & Coyne 1974).

The symptoms produced in susceptible plants by bacterial pathogens are varied and include leaf spots and streaks, branch cankers, total wilt, soft rots of stems or storage organs (see Chs 4 & 5, this volume) and the production of galls.

In this paper some of the bacterial diseases of food plants are discussed with a view to providing insight into some of the problems facing the plant bacteriologist, to indicate how they are being overcome, and to consider those fields in which future research is most urgently needed.

The more important botanical families attacked by bacteria are the Gramineae (constituting *ca.* 75% of the world's food supply), Solanaceae, Leguminosae and Rosaceae. The most important world crops placed in order of tonnage produced are listed in Table 1.

Some crops, e.g. tomato, cassava, rice and beans, are susceptible to three or more bacterial pathogens, many crops have only one, and a few,

TABLE 1
Some of the more important world food crops

In order of tonnage produced	Crop	Metric tonnes ($\times 10^6$)	Main production area (and its % of the world crop)
1	Cane sugar	781	Asia 42
2	Wheat	441	USSR 27
3	Rice	376	Asia 91
4	Maize	363	N. America 54
5	Beet sugar	289	Europe 49
6	Potato	273	Europe 44
7	Barley	196	Europe 36
8	Cassava	119	Africa 16
9	Sweet potato	99	Asia 90
10	Sorghum	69	Asia 35
11	Soybean	80	N. America 54
12	Banana	57	Africa 31
13	Oats	50	USSR 37
14	Citrus fruits	46	N. America 27
15	Tomato	47	Europe 29

Data from FAO Yearbook (1978).

e.g. cocoa, have no known bacterial pathogen. At one extreme are organisms with a very wide host range; *Agrobacterium tumefaciens*, for example, can infect over 600 plant species: *Pseudomonas solanacearum* and *Ps. syringae* also attack a variety of hosts. However, most pathogens are restricted to one or a few plant species while others even discriminate between varieties of the same species.

2. Isolation and Detection of Bacteria

Some of the problems encountered in the isolation of plant pathogenic bacteria arise from the general pathologist's lack of appreciation that the techniques are different from those used in investigation of fungal pathogens, a deficiency Bradbury (1970) has attempted to meet. Isolation is also hampered by a lack of good selective media. However, with experience in the selection of tissue at an early stage of infection and in colony recognition, then a simple nutrient agar is often adequate. For samples heavily contaminated with soil or secondary organisms, media with at least some degree of selectivity are required.

When a pathogen is present in small numbers (in seeds, dormant buds etc.) and isolation is virtually impossible, presumptive evidence of its presence may be obtained from the population increase of a highly specific phage (Billing & Garrett 1980) or from fluorescent pigment production, e.g. from infected beans (Wharton 1967). Serological techniques can also be used to detect bacterial pathogens in soil and plant tissues (see p.119) avoiding the immediate need to isolate the pathogen before positively identifying it.

3. Identification Methods

Accurate identification of a pathogen is a necessary prerequisite to control. Most bacterial pathogens can readily be assigned to their correct genera but the nomenspecies or pathovar is more difficult to determine. Facilities for the more sophisticated modern methods of DNA composition and homology, cell wall and lipid analyses, and numerical taxonomy, are not generally available to the field bacteriologist, although their use will undoubtedly clarify current taxonomic problems.

A. Pathogenicity tests

Plant inoculation to reproduce field symptoms is best used when the bacterium is recovered in nearly pure culture, when suitable host tissue is readily available, and space and time are not at a premium. Positive hypersensitivity reactions (HR) in tobacco (Klement & Goodman 1967) and lesion production on bean pods or immature fruits, using high inoculum doses, confirm pathogenicity but do not necessarily indicate ability to establish disease in the field on a particular host.

B. Cultural, biochemical and physiological tests

Sometimes conventional cultural, biochemical and physiological tests are sufficient to identify a pathogen. Often they are time consuming and fail to distinguish between closely related organisms, whether pathogens or saprophytes, perhaps because many tests, e.g. gelatin liquefaction, bear no relation to the plant habitat of the pathogens. For many plant pathologists, however, these tests are the only available means of identification.

An investigation into the value and reliability of these tests for pseudomonads showed that although a few, e.g. arginine hydrolysis, gave highly reproducible results, some, e.g. ability to grow at 37°C, were very unsatisfactory. Experience of workers, time of recording and inherent test variation were all important variables but differences in brand of product or type of container used were of little consequence (Sneath & Collins 1974). In a later international study of plant pathogenic pseudomonads several selected tests were either not reproducible or had no differential value (Lelliott 1978). These studies are a salutory warning against drawing hasty conclusions about the identity of new isolates with published descriptions of pathogens on the basis of a limited number of isolates or tests.

C. Phage and bacteriocin typing

Phage typing, commonly used for some animal pathogens, e.g. *Salmonella* spp., has had little impact on the identification of plant pathogenic bacteria (Billing & Garrett 1980). Highly specific phages have, however, been used in epidemiological studies to identify organisms at the subspecific level (Mizukami & Wakimoto 1969; Crosse & Garrett 1970).

The bacteriocins of plant pathogenic bacteria have been little studied although a range of them are now known to exist (Vidaver 1976). For some organisms, where conventional identification tests seem inadequate, e.g. for the coryneforms, the development of phage or bacteriocin typing schemes might prove valuable (Vidaver 1976; Billing & Garrett 1980).

D. Serology

The use of serological methods in the identification of plant pathogenic bacteria is rapidly expanding, partly due to improved techniques in the preparation of antisera (Trigalet *et al.* 1978; Schaad 1979). These methods are generally very sensitive, more or less specific and relatively easy to perform though some require special equipment and expertise. Several workers are now using immunofluorescence routinely to identify the pathogens *Erwinia carotovora* var. *atroseptica*, *Xanthomonas campestris* and *Pseudomonas solanacearum* (Coleno *et al.* 1976; Trigalet *et al.* 1978; Schaad 1979).

The many contrasting results in the literature indicate that these serological tests are not, as yet, wholly reliable. Trigalet *et al.* (1978) and Schaad (1979) emphasize the need for careful preparation of every reagent and accurate evaluation of its degree of specificity. They hope that, ultimately, universally standardized procedures will be devised and adopted to increase reliability, and commend these rapid and valuable tests to plant bacteriologists.

E. Choice of method

The widespread exchange of plant material and its fast intercontinental transport make imperative the use of rapid and reliable tests for identification of plant pathogenic bacteria. The pathologist must select those tests for which the material, facilities and expertise are available, and that give the sensitivity required. Until more selective media are developed, more tests determining the utilization of compounds of plant origin, rather than animal origin, are needed. Guidelines for the standardization of tests (pathogenicity, biochemical, phage, bacteriocin and serological) should be compiled to increase their reliability and differential value; meanwhile, identification of phytopathogenic bacteria will continue to be a problem. It is perhaps in the development of serological methods that the greatest promise lies.

4. Economic Importance

A. General considerations

Agriculture involves a constant adjustment between the crop and its competitors whether they are weeds, insects or micro-organisms. Over 30% of the world's crops are lost in the field and a further 15% is destroyed during transit and storage, despite the widespread use of pesticides. Diseases, of which those of known bacterial origin are but a small part, account for an estimated 12–18% of the world-crop loss (Cramer 1967).

Plant diseases were originally studied because of their destruction of crops. The importance of a pathogen is measured by estimates of the loss of crop it causes, yet, even today, statements such as "causes serious loss" occur with greater frequency than accurate data. Loss of crop, which is difficult both to define and quantify (Large 1966), is the subject of a specialized and expanding field of investigation because of the economic necessity to be better informed about the extent and nature of loss, either to justify the use of expensive chemical controls or to deploy aid to developing countries to the best advantage (Ordish & Dufour 1969; Carlson & Main 1976; James & Teng 1979). One outcome of an FAO symposium on crop loss assessment in 1967 was the production of a manual of methods to assess losses caused by specified organisms (Chiarappa 1971). Significantly, no bacteria are among the 130 organisms cited in the manual and its two supplements. An international committee of the International Society of Plant Pathologists (ISPP) is currently helping FAO to prepare an up-to-date crop loss manual. It is to be hoped that bacterial diseases will not be neglected, for although fungal diseases are of great economic importance, bacterial plant pathogens can cause serious crop losses as the following selected examples indicate.

B. *Some important bacterial diseases*

(i) *Bacterial wilt of Solanaceae*

Bacterial wilt (*Pseudomonas solanacearum*), the most widespread and serious bacterial disease of plants, is the subject of *ca.* 1500 papers of which the reviews by Kelman (1953) and Buddenhagen & Kelman (1964) are outstanding. The disease is largely confined within the 45°N and 45°S latitudes, where annual rainfall exceeds 100 cm, winter and summer temperatures are over 10°C and 20°C, respectively, and the growing season is longer than 6 months. Infected plants die prematurely, resulting in decreased yields and decay during storage. Crop losses of from 50–100% have been recorded, particularly for potato but also for tomato, eggplant and banana in areas of Africa, India and Indonesia (Kelman 1953). The wide variation in the characteristics of isolates of the pathogen has resulted in several subspecific classifications, based on different criteria, and this is a major problem which has discouraged research workers from the all-important identification of strains which is necessary for determination of varietal resistance and disease control. More needs to be known about the ecology of the different strains and their potential as pathogens (Kelman 1976). Three races of the pathogen are recognized, within each of which are a number of pathotypes. Race 1 attacks solanaceous and other hosts; race 2 attacks bananas and *Heliconia* spp.; race 3 attacks potato, and has a narrow host range and a lower optimum temperature for growth than the others.

Effective controls are lacking: the use of chemicals and antibiotics is often impracticable in peasant economies. The very wide host range of the pathogen and the fact that it is a soil inhabitant make cultural methods of control very difficult.

Development of resistant varieties is the only real solution to the control of bacterial wilt (see p.128) but the plant breeders' problems are accentuated by the wide strain variation in this pathogen.

(ii) *Leaf blight of rice*

Most of the world's rice, the staple food for 60% of humanity, is produced in Asia (Table 1). Bacterial leaf blight caused by *Xanthomonas oryzae*, first recognized in Japan in 1884, has become of major importance since the 1950s following the introduction of high-yielding, short, stiff-strawed varieties which were disastrously susceptible to leaf blight (Ou 1972). The more virulent strains of the pathogen cause the 'Kresek' form of the disease which commonly causes losses of up to 60% in India and

Indonesia. Cultural methods, e.g. irrigation and flooding of fields, together with heavy rains and typhoons are major disseminators of the pathogen which survives the year round on the rice crop, on stubble or even on weeds on the banks of irrigation channels (Mizukami & Wakimoto 1969). Infected seed is the most important inoculum source in India (Srivastava & Rao 1964) and is thought to have been responsible for the recent introduction of the disease into Central and South America.

(iii) *Bacterial canker of stone fruits*

Fluorescent pseudomonads of the *Ps. syringae* group cause canker diseases on stone fruits in many countries but losses are often difficult to estimate. However, the recent occurrence on peach trees of a distinct pathogen closely related to other stone fruit pathotypes (Crosse 1968) provided an opportunity to record the spread of this particularly destructive disease and the losses it caused. The die-back and death of peach trees, first noted on the plateau above the Rhône valley of France in 1966, is caused by the bacterium *Ps. persicae* (Luisetti *et al.* 1976). The disease has already destroyed *ca.* 750,000 peach trees in the Ardèche and neighbouring *départements* (Prunier *et al.* 1976; Prunier, J-P., pers. comm.). There was a three-fold increase in tree deaths from canker in 1975 following the severe spring frosts which increased the susceptibility of peach to the pathogen. Death of mature trees means loss of crop not only in that year but in all the succeeding years until replacements are in full bearing. In this intensive peach production area, mainly in small farm units with the growers almost totally dependent on the one crop, the socio-economic consequences of such a disease are considerable.

(iv) *Fireblight and other diseases in the USA*

Since the discovery in the 1880s that fireblight was caused by a bacterium, millions of apple and pear trees in the USA have been destroyed by *Erwinia amylovora* and pear production in some states has been abandoned; in others control measures are expensive and not very effective (van der Zwet & Keil 1979).

Although fireblight remains a serious problem and is ranked among the three most important bacterial diseases in 17 States of the USA (Table 2) a 1976 survey there indicated that losses from *Ps. glycinea* on soybeans and from crown gall on fruit and nuts were much greater. Of the 38 bacterial pathogens covered in the survey, six accounted for 75% of the losses caused (Table 2) (Kennedy & Alcorn 1980).

It is not possible to mention many bacterial diseases here but the above

TABLE 2
Crop losses from bacterial plant pathogens in USA

Bacterium	Crop	No. States placing as in top 3 diseases	Estimated loss ($ M)
Pseudomonas glycinea	soybeans	8	65
Agrobacterium tumefaciens	fruit, nuts	10	23
Pseudomonas syringae	wheat	13	18
Erwinia carotovora	potatoes	16	14
Xanthomonas phaseoli	beans	12	5
Erwinia amylovora	pears	17	4·7
			129·7
32 other bacterial spp.			45·43
			175·13

Data from Kennedy & Alcorn (1980).

examples give some indication of their economic importance, the roles played by environmental factors and by alternative hosts, our need for greater understanding of the strain specificity of the pathogens, and the difficulties of disease control.

5. Control Measures

Much can be done to reduce the attacks of bacterial pathogens by good husbandry—choice of site, spacing and varieties of the crop plant, careful selection of manurial programmes and pruning systems, and attention to time of sowing, crop rotation and irrigation methods. In developing countries growers often have recourse to nothing else. As a complement to good husbandry, but not in place of it, growers in the developed countries should resort to other control measures, otherwise their large areas of crop monoculture, which create ideal epidemic conditions, may be devastated by disease. Unfortunately, bacterial diseases of plants remain notoriously difficult to control and continue to take their toll. The inadequacy of existing controls, coupled with environmental and economic considerations, has stimulated research into measures other than the use of broad-spectrum chemicals and antibiotics. But if maximum benefit is to be derived, more research into the epidemiology of the diseases and the ecology of the causal pathogens is needed so that effective measures can be formulated within the inevitable limitations of what is practicable and economic for the grower.

A. Exclusion and eradication

The widespread exchange of plant material between countries and continents today increases the risk of diseases being introduced into new areas. Legislation limits the movement of plant material, seeds being one of the commonest sources of primary infection and from which secondary spread can soon result in epidemics. As few as two infected seeds per kilogram can cause serious outbreaks of halo blight (caused by *Ps. phaseolicola*) on field beans: only one seed infected with *X. campestris* in 10,000 is tolerated for heading cabbage; for seed production no infection is tolerated (Schaad *et al.* 1980). The importation into UK of tomato or pea seed is prohibited to exclude *Corynebacterium michiganense* and *Ps. pisi*, respectively. The many facets of administration to maintain plant health were discussed at a symposium in 1978 (Ebbels & King 1979).

Seed treatments can eliminate the pathogen but are often not wholly effective. Hot water treatment has been used against *X. campestris* in brassica, *Ps. lachrymans* in cucumber and *C. michiganense* in tomato seed. *Pseudomonas tomato* infection can be removed by acid or fermentation extraction methods. Bean seeds can be freed from the pathogens attacking them by treatment with streptomycin slurries (Taylor & Dye 1972). In spite of legislation and treatment of material, new introductions of pathogens sometimes occur although by the time the disease is noticed the pathogen population has increased and the chances of eliminating it have thereby decreased. Nevertheless, in a restricted area eradication may be attempted where there is reasonable hope of success and the long-term benefits of so doing outweigh any high cost involved.

The classic example is the costly eradication of *X. citri* from the USA (Dopson 1964). An industry-supported programme to eliminate *Ps. solanacearum* from seed potatoes in Victoria, Australia is reported to have eradicated wilt from areas of recent introduction of the pathogen, and controlled its spread into others (Harrison 1976).

Eradication programmes are not always so successful. In spite of determined and strenuous efforts to eradicate fireblight (*E. amylovora*) following its introduction into England, the Netherlands and France between 1958–1972, the campaigns had to be abandoned because early spread into woodland and ornamental hosts ensured a reservoir of infection (Lelliott 1968; van der Zwet & Keil 1979). Nevertheless, the disease was contained for several years but it is now established in N. Europe and threatens fruit production in the warmer Mediterranean areas.

B. Chemical control

Since the discovery of Bordeaux mixture in 1882, the chemical control of plant diseases has been practised routinely, but among the vast and ever-increasing number of pesticides manufactured today there are few bactericides. Crosse (1971*b*) stressed the need for systemic materials less phytotoxic than copper, but no real advances towards this goal have been made. Therefore the surface-acting formulations of copper etc. continue to be widely used against a range of bacterial diseases (Crosse 1971*b*). Adjuvants such as vegetable oils reduce the phytotoxic effects of these sprays.

C. Antibiotics

Streptomycin has been widely and successfully used against *E. amylovora* fireblight in New Zealand and the USA since the 1950s because it gives better control than other bactericides and is non-phytotoxic (van der Zwet & Keil 1979). However, because streptomycin-resistant strains of *E. amylovora* now occur in over 40% of the pear acreage of California, where sprays have been applied very frequently, terramycin must now be used (van der Zwet & Keil 1979). Streptomycin control of *X. vesicatoria* leaf spot of pepper also lost effectiveness when drug resistant strains emerged (Thayer & Stall 1961).

Where control by streptomycin is no better than copper sprays against *Ps. phaseolicola* halo blight, *Ps. syringae* bacterial canker and *Ps. tomato* tomato speck, its use is uneconomic. Sometimes an antibiotic gives superior control, e.g. oxytetracycline against *Ps. persicae* peach canker in France, but its commercial use is prohibited (Prunier *et al.* 1976). Although antibiotics have the advantage of being locally systemic, their use to control plant diseases is by no means universally approved and in some, including most European, countries is prohibited. The agricultural use of any antibiotic having medicinal application should be discouraged because of the potential hazard of transfer of drug resistance from bacteria associated with plants to those associated with animals and humans (Byrde *et al.* 1979).

D. Biological control

Among the complex populations of micro-organisms on, or around, plants are some that are antagonistic to bacterial plant pathogens. If these antagonists can be introduced, or the local environment modified in their favour then biological control could prove effective, selective, economic and environmentally acceptable. Difficulties of preparation and storage of antagonists need to be overcome and the dangers of organisms acting as genetic vectors of resistance need to be recognized.

(i) *Phages*

It is doubtful whether phages contribute significantly to the natural regulation of plant pathogenic bacteria. Hopes that they might be a useful specific control agent of plant diseases have not been fulfilled. Even if a high enough phage population could be maintained for efficient contact with, and thereby control of, the pathogen, the attendant risks of altering the genetic status of the phages, the bacterial pathogens or even the host plants, are considered too great to permit their use (Vidaver 1976).

(ii) *Bacteriocins*

An exciting breakthrough in the last decade has brought about the first successful biological control of a bacterial plant pathogen. Crown gall of peach, caused by *Agrobacterium tumefaciens*, is now controlled by an antagonist, *A. radiobacter* strain 84, which produces the bacteriocin agrocin 84. Kerr (1980) has reviewed the development of this preventive control from his discovery of strain 84 to our present understanding of its mode of action (see Ch. 7, this volume).

The rapid and widespread commercial adoption of strain 84 reflects a lack of alternative control measures, and is not always appropriate because some pathogenic isolates of *A. tumefaciens*, especially those from grapevine, are insensitive to agrocin 84. Bacteriocins of other strains of *A. radiobacter* have so far failed to control agrocin-84-resistant strains of the pathogen in the field (Moore & Warren 1979). It may be that these bacteriocins are produced in insufficient quantities at the site of action to be effective.

Other bacteriocins have been examined of which the syringacins are reported to suppress lesion development by *Ps. phaseolicola* on bean leaves and by *Ps. glycinea* on bean seeds (Vidaver 1976).

(iii) *Other antagonists*

Since Crosse (1971*a*) reviewed the interactions between plant pathogenic and other organisms in the plant environment, research has continued but has not yet led to acceptable commercial control. Progress towards biological control of fireblight has been slow although the use of an *Erwinia* sp., bacteriocin-producing strains of *E. herbicola* and avirulent *E.*

amylovora and *Ps. syringae* have shown some promise (Aldwinckle & Beer 1978; Paulin & Lachaud 1978).

The best prospects for biological control of pathogens lie in selectively enhancing growth and activity of the antagonistic components of the normal microflora of aerial parts or rhizosphere, although, as many have discovered, promising *in vitro* activity is frequently not matched *in vivo*.

E. Disease forecasting to aid control

Accurate forecasting to evaluate the probability of infection, which would enable the use of effective control measures at the optimum time to reduce or entirely prevent disease can help to minimize the use of expensive and polluting pesticides. Temperature and precipitation, whether as rainfall or dew, are the main factors used in most disease forecasting systems, particularly when wind-driven rain is important in the secondary spread of a pathogen.

Many of the past difficulties of weather and biological data collection and collation have now been overcome by the use of highly sensitive electronic devices, statistical methods and computer analysis, and it is probable that more, and increasingly accurate, disease forecasting systems will be devised for a range of crops (Krause & Massie 1975).

Predictive systems for fireblight, based on temperature thresholds and precipitation, have been used extensively in the USA since the 1950s and in California alone are reported to have saved $0·75 million in spray costs in 1975 and 1976 (Aldwinckle & Beer 1978). These systems have been compared with Billing's system in the UK—the first that took into account *in vitro* studies of the effect of temperature on growth of the pathogen, *E. amylovora* (Billing 1980). Billing suggested a graded system for assessing the risk of fireblight but it needs to be examined over many years and in many localities before its precision can be evaluated fully.

Another approach to disease forecasting is the monitoring of the build-up of specific *X. oryzae* phage in paddy field irrigation channels to assess population levels of host *X. oryzae*. Although abnormal climatic conditions can upset the forecasts they have proved valuable in aiding chemical control of leaf blight of rice (Mizukami & Wakimoto 1969).

A more recent innovation concerning disease control is that of aerial surveys of field beans, whereby infection foci can be detected before symptoms are visible to ground level observers. When problems of data interpretation can be overcome, the potential application of such methods to aid optimum timing of sprays can be assessed (Wallen & Galway 1979) in relation to their cost effectiveness.

F. Breeding resistant varieties

Breeding disease-resistant varieties to improve yields and reduce costs is the best long-term solution to minimize crop loss from disease. Resistant varieties have been developed against all major parasitic groups of organisms, including bacteria, but less is known about the nature of resistance to bacteria than to fungi (Russell 1978). Development of new varieties, especially of perennial crops, is time consuming and expensive. The limiting factors in the use of resistant varieties include the degree of host resistance that can be achieved and its durability to the evolving pathogen. Techniques for the initial screening of seedlings, especially of perennial crops, must be established to ensure that they predict accurately subsequent field performance (Mizukami & Wakimoto 1969; van der Zwet & Keil 1979; Garrett 1979).

For a number of crop plants varieties resistant to their bacterial pathogens exist. Resistance to *Ps. solanacearum* was bred in the Schwarz 21 line of peanut over 50 years ago and it is still widely planted and used as a parent in breeding programmes. Bacterial (*Ps. solanacearum*) wilt-resistant varieties of eggplant and pepper are available but success with other hosts has been moderate. In tomato, *Ps. solanacearum* resistance is subject to breakdown at high temperatures, and in potato, where resistance levels are not very high, as many as 10 major genes are thought to be involved.

Resistant varieties of rice are widely grown but they are not immune to the most virulent strains of *X. oryzae* which can attack the crop under a variety of conditions (Mizukami & Wakimoto 1969; Ou 1972).

Apple- and pear-breeding programmes were begun in the USA over a century ago, but although a great range of new varieties has been released, production of good quality fruit from trees highly resistant to *E. amylovora* fireblight has so far eluded the breeders (van der Zwet & Keil 1979).

Sometimes the early promise of new varieties from breeding programmes is not realized. For example *X. oryzae* became pandemic in Asia only after the introduction of new high yielding rice varieties (Ou 1972); mosaic-tolerant varieties of cassava were decimated by bacterial blight, caused by *X. cassavae*, in West Africa (Persley 1976); and an epiphytotic of *X. phaseoli* bacterial blight on beans in Canada coincided with the planting of an anthracnose-resistant variety (Wallen & Galway 1979). Thus, new varieties may be susceptible to a minor pathogen to which indigenous or long-established varieties are resistant, and attempts to improve crops may unwittingly increase the importance of some diseases. Greater co-operation between plant breeders and pathologists of different disciplines is essential if disasters are to be avoided.

Breeding of crop plants must be a continuing process because of variability in the pathogens and the changing requirements of the food industry (for varieties suitable for canning, freezing, mechanical harvesting etc.). The new varieties must not only be disease resistant, but also of superior quality to overcome the growers' natural reluctance to change from a favourite, albeit disease-susceptible, variety.

6. Conclusions

On a world-wide scale the incidence and importance of known bacterial diseases of food crops, and the number of workers involved in their study, is small. Nevertheless some bacterial diseases contribute significantly to food production losses, as may become more evident when accurate crop loss assessment methods are devised and used. More reliable tests for identifying pathogens will speed their recognition and help pathologists and growers to take appropriate action more quickly.

None of the sessions in the international conferences on plant pathogenic bacteria in 1971 and 1978 was directly concerned with the control of bacterial diseases other than biological control of crown gall caused by *A. tumefaciens*. This reflects a continuing lack of field-effective bactericides and failure to develop alternative and more imaginative methods of control, rather than disinterest in the solution of disease control problems. The scale of losses from bacterial diseases, although critical for certain crops, is apparently too small to justify to pesticide manufacturers the developmental costs for an agricultural bactericide. Chemicals developed for medical or veterinary use have generally proved ineffective or environmentally unacceptable for use on crops. Such considerations necessitate research to improve the efficiency of existing control methods and to search for alternative measures by detailed investigations of the epidemiology, ecology and host–parasite relations of the pathogens. Success in this search is most likely to result from an integrated programme of research into cultural practices, breeding for resistant varieties, and developing chemical and biological control methods.

Many plant bacteriologists are working in isolation. International co-operation, conferences and workshops are vital for the exchange of ideas and dissemination of information. All these have increased in recent years and should lead to greater co-ordination of the limited, and often scattered and fragmentary, research effort into bacterial diseases of plants.

7. References

Aldwinckle, H. S. & Beer, S. V. 1978 Fireblight and its control. *Horticultural Reviews* **1**, 423–474.

Billing, E. 1980 Fireblight (*Erwinia amylovora*) and weather: a comparison of warning systems. *Annals of Applied Biology* **95**, 365–377.

Billing, E. & Garrett, C. M. E. 1980 Phages in the differentiation of Plant Pathogenic Bacteria. In *Microbiological Identification* ed. Goodfellow, M. & Board, R. G. pp.319–338. Society for Applied Bacteriology Symposium Series No. 8. London & New York: Academic Press.

Bradbury, J. F. 1970 Isolation and preliminary study of bacteria from plants. *Review of Plant Pathology* **49**, 213–218.

Breed, R. S., Murray, E. G. D. & Smith, N. R. eds 1957 *Bergey's Manual of Determinative Bacteriology* 7th edn. Baltimore: Williams & Wilkins.

Buchanan, R. E. & Gibbons, N. E. eds 1974 *Bergey's Manual of Determinative Bacteriology* 8th edn. Baltimore: Williams & Wilkins.

Buddenhagen, I. & Kelman, A. 1964 Biological and physiological aspects of bacterial wilt caused by *Pseudomonas solanacearum*. *Annual Review of Phytopathology* **2**, 203–230.

Byrde, R. J. W. *et al.* 1979 Problems and prospects of chemical control of plant diseases. *Committee on Chemical Control ISPP* Special Report No. 1.

Carlson, G. A. & Main, C. E. 1976 Economics of disease-loss management. *Annual Review of Phytopathology* **14**, 381–403.

Chiarappa, L. ed. 1971 *Crop Loss Assessment Methods.* (FAO Manual on the evaluation and prevention of losses by pests, disease and weeds.) FAO Publication of United Nations by CAB.

Coleno, A., Trigalet, A. & Digat, B. 1976 Détection des lots de semences contaminés par une bacterié phytopathogène. *Annales Phytopathologie* **8**, 355–364.

Cramer, H. H. 1967 *Plant Protection and World Crop Production* p.524. Leverkusen: Bayer.

Crosse, J. E. 1957 The dispersal of bacterial plant pathogens. In *Biological Aspects of the Transmission of Disease* pp.7–12. Edinburgh: Oliver & Boyd.

Crosse, J. E. 1968 Plant pathogenic bacteria in soil. In *The Ecology of Soil Bacteria* ed. Gray, T. R. B. & Parkinson, D. pp.552–572. Liverpool University Press.

Crosse, J. E. 1971*a* Interactions between saprophytic and pathogenic bacteria in plant disease. In *Ecology of Leaf Surface Microorganisms* ed. Preece, T. F. & Dickinson, C. H. pp.283–290. London & New York: Academic Press.

Crosse, J. E. 1971*b* Prospects for the use of bactericides for the control of bacterial diseases. In *Proceedings 6th British Insecticide & Fungicide Conference* pp.694–705.

Crosse, J. E. & Garrett, C. M. E. 1970 Pathogenicity of *Pseudomonas morsprunorum* in relation to host specificity. *Journal of General Microbiology* **62**, 315–327.

Dopson, R. N. 1964 The eradication of citrus canker. *Plant Disease Reporter* **48**, 30–31.

Dye, D. W., Bradbury, J. F., Dickey, R. S., Goto, M., Hale, C. N., Hayward, A. C., Kelman, A., Lelliott, R. A., Patel, P. N., Sands, D. C., Schroth, M. N., Watson, D. R. W. & Young, J. M. 1975 Proposals for a reappraisal of the status of the names of plant-pathogenic *Pseudomonas* species. *International Journal of Systematic Bacteriology* **25**, 252–257.

Ebbels, D. L. & King, J. E. eds 1979 *Plant Health. The Scientific Basis for Administrative Control of Plant Diseases and Pests* p.322. Oxford: Blackwell Scientific.

FAO 1979 FAO Production Yearbook for 1978 Vol. 32. FAO Statistics Series No. 22. FAO, Rome.

Garrett, C. M. E. 1979 Screening *Prunus* rootstocks for resistance to bacterial canker, caused by *Pseudomonas morsprunorum*. *Journal of Horticultural Science* **54**, 189–193.

Harrison, D. E. 1976 Control of bacterial wilt of potatoes in Victoria, Australia by exclusion. In *Proceedings First International Planning Conference and Workshop on the Ecology and Control of Bacterial Wilt caused by* Pseudomonas solanacearum ed. Sequeira, L. & Kelman, A. Raleigh: N. Carolina State University.

James, W. C. & Teng, P. S. 1979 The quantification of production constraints associated with plant diseases. *Applied Biology* **4**, 201–267.

KELMAN, A. 1953 The bacterial wilt caused by *Pseudomonas solanacearum. Technical Bulletin 99, North Carolina Agricultural Experiment Station* pp.194.

KELMAN, A. 1976 Mission of the Conference. In *Proceedings First International Planning Conference and Workshop on the Ecology and Control of Bacterial Wilt caused by* Pseudomonas solanacearum ed. Sequeira, L. & Kelman, A. Raleigh: N. Carolina State University.

KENNEDY, B. W. & ALCORN, S. M. 1980 Estimates of U.S. crop losses due to procaryote plant pathogens. *Plant Disease* **64**, 674–676.

KERR, A. 1980 Biological control of crown gall through production of agrocin 84. *Plant Disease* **64**, 25–30.

KLEMENT, Z. & GOODMAN, R. N. 1967 The hypersensitive reaction to infection by bacterial plant pathogens. *Annual Review of Phytopathology* **5**, 17–44.

KRAUSE, R. A. & MASSIE, L. B. 1975 Predictive systems: modern approaches to disease control. *Annual Review of Phytopathology* **13**, 31–47.

LARGE, E. C. 1966 Measuring plant disease. *Annual Review of Phytopathology* **4**, 9–28.

LELLIOTT, R. A. 1968 Fireblight in England, its nature and its attempted eradication. *EPPO Report of the International Conference on Fireblight*, Series A No. 45E, pp.10–14.

LELLIOTT, R. A. 1978 Report on the Workshop of the 1978 International Collaborative Study of *Pseudomonas* group. *Proceedings 4th International Conference on Plant Phytopathogenic Bacteria* Angers, 1978, p.735.

LUISETTI, J., PRUNIER, J-P., GARDAN, L., GAIGNARD, J-L. & VIGOUROUX, A. 1976 *Le Dépérissement Bactérien du Pêcher* (Pseudomonas morsprunorum *f. sp.* persicae). Paris: INVUFLEC.

MIZUKAMI, T. & WAKIMOTO, S. 1969 Epidemiology and control of bacterial leaf blight of rice. *Annual Review of Phytopathology* **7**, 51–72.

MOORE, L. W. & WARREN, G. 1979 *Agrobacterium radiobacter* strain 84 and biological control of crown gall. *Annual Review of Phytopathology* **17**, 163–179.

ORDISH, G. & DUFOUR, D. 1969 Economic bases for protection against plant diseases. *Annual Review of Phytopathology* **7**, 31–50.

OU, S. H. 1972 *Rice Diseases* p.368. Kew: Commonwealth Mycological Institute.

PAULIN, J-P. & LACHAUD, G. 1978 Fireblight situation in France: August 78. *Proceedings 4th International Conference on Phytopathogenic Bacteria* Angers, 1978, pp.519–520.

PERSLEY, G. J. 1976 *Report on Cassava Disease Survey in Cameroon, 5–20 June, 1976* p.30. Ibadan: IITA.

PRUNIER, J-P., GAIGNARD, J-L., GARDAN, L., LUISETTI, J. & VIGOUROUX, A. 1976 Le dépérissement bactérien du pêcher dans le sud-est de la France. *Arboriculture Fruitière* **23**, 21–24.

RUSSELL, G. E. 1978 *Plant Breeding for Pest and Disease Resistance* p.485. London: Butterworth.

SCHAAD, N. W. 1979 Serological identification of plant pathogenic bacteria. *Annual Review of Phytopathology* **17**, 123–147.

SCHAAD, N. W., SITTERLY, W. R. & HUMAYDAN, H. 1980 Relationship of incidence of seedborne *Xanthomonas campestris* to black rot of crucifers. *Plant Disease* **64**, 91–92.

SCHUSTER, M. L. & COYNE, D. P. 1974 Survival mechanisms of phytopathogenic bacteria. *Annual Review of Phytopathology* **12**, 199–221.

SNEATH, P. H. A. & COLLINS, V. 1974 A study in test reproducibility between laboratories: Report of a *Pseudomonas* working party. *Antonie van Leeuwenhoek* **40**, 481–527.

SRIVASTAVA, D. N. & RAO, Y. P. 1964 Seed transmission and epidemiology of the bacterial blight disease of rice in North India. *Indian Phytopathology* **17**, 77–78.

TAYLOR, J. D. & DYE, D. W. 1972 A survey of the organisms associated with bacterial blight of peas. *New Zealand Journal of Agricultural Research* **15**, 432.

THAYER, P. L. & STALL, R. E. 1961 Effect of variation in the bacterial spot pathogen of pepper and tomato on control with streptomycin. *Phytopathology* **51**, 568–571.

THURSTON, H. D. 1973 Threatening plant diseases. *Annual Review of Phytopathology* **11**, 27–52.

TRIGALET, A., SAMSON, R. & COLENO, A. 1978 Problems related to the use of serology in phytopathology. *Proceedings 4th International Conference on Phytopathogenic Bacteria* Angers, 1978, pp.271–288.

van der Zwet, T. & Keil, H. L. 1979 Fireblight a bacterial disease of Rosaceous plants. In *USDA Agriculture Handbook* No. 510, p.200.

Vidaver, A. K. 1976 Prospects for the control of phytopathogenic bacteria by bacteriophages and bacteriocins. *Annual Review of Phytopathology* **14**, 451–465.

Wallen, V. R. & Galway, D. A. 1979 Effective management of bacterial blight of field beans in Ontario — a 10-year program. *Canadian Journal of Plant Pathology* **1**, 42–46.

Wharton, A. L. 1967 Detection of infection by *Pseudomonas phaseolicola* (Burkh.) Dowson in white seeded dwarf bean stocks. *Annals of Applied Biology* **60**, 305–312.

Young, J. M., Dye, D. W., Bradbury, J. F., Panagopoulos, C. G. & Robbs, C. F. 1978 A proposed nomenclature and classification for plant pathogenic bacteria. *New Zealand Journal of Agricultural Research* **21**, 153–177.

All *Pseudomonas* and *Xanthomonas* species cited in this paper, with the single exception of *Ps. solanacearum*, are now more properly designated *Ps. syringae* and *X. campestris* with the earlier specific epithet reduced to the status of pathovar within these two heterogeneous species (see Dye, D. W., Bradbury, J. F., Goto, M., Hayward, A. C., Lelliot, R. A. & Schroth, M. N. International standards for naming pathovars of phytopathogenic bacteria and a list of pathovar names and pathotype strains. *Review of Plant Pathology* **59**, 153–168, 1980).

The Effect of Bacteria on Post-harvest Quality of Vegetables and Fruits, with Particular Reference to Spoilage

BARBARA M. LUND

Agricultural Research Council, Food Research Institute, Norwich, UK

Contents

1. Introduction

THE PURPOSE of this paper is to outline the main aspects of the subject; the references are not exhaustive but are intended to provide an entry into the literature on each topic.

Considerable wastage of fresh fruit and vegetables occurs between the farm and the consumer. Losses have been estimated at about 12% of total production in the USA, while in tropical regions they are likely to be much greater (Coursey & Booth 1972; Eckert 1975; Harvey 1978). Such post-harvest losses may wipe out a large investment in growth, harvesting, packaging, storage and transportation of a crop (Eckert 1978). This wastage is caused by three factors: mechanical damage, physiological changes including loss of water and consequent wilting, and the effect of spoilage micro-organisms.

With certain exceptions, e.g. cucumbers, peppers and tomatoes, microbial spoilage of fruits is usually caused by fungi rather than by bacteria, this is probably due in part to the fact that the pH value of fruits (usually <4·5, von Schelhorn 1951) inhibits growth of most spoilage bacteria. The higher pH values within vegetables (4·5–7·0) allow growth of bacteria, which cause a significant proportion of the microbial spoilage. The most recent and accurate estimates of post-harvest losses are those reported for selected types of fruit and vegetables sampled in the New York and Chicago areas at wholesale, retail and consumer levels (Harvey 1978). Bacterial soft rot was a major cause of microbial spoilage of lettuce (Ceponis 1970; Ceponis *et al.* 1970; Beraha & Kwolek 1975), potatoes (Ceponis & Butterfield 1973), bell peppers (Ceponis & Butterfield 1974*a*), cucumbers (Ceponis & Butterfield 1974*b*) and tomatoes (Ceponis & Butterfield 1979).

Post-harvest spoilage can be caused by bacteria which are present on vegetables or fruit in the field, or with which the produce is contaminated at harvest or during subsequent handling, particularly washing. Bacterial attack may have started in the field and there is no clear distinction between field disease and post-harvest spoilage. Whether in the field or after harvest, bacteria may enter the plant tissue through natural openings, e.g. hydathodes, stomata and lenticels (Ch. 4, this volume) or through tissue damaged by machinery, fungi, nematodes, small mammals or birds, or by rain and hail.

The most economically important bacterial spoilage is probably that caused by soft rots, characterized by disintegration of the plant tissue, but bacteria also cause other defects which lower the quality of the produce and allow soft rot bacteria to enter.

2. Defects Caused by Bacteria

Where a reference is cited reporting disease or spoilage the bacterial nomenclature used by the author is quoted. In other cases the nomenclature used is that given in the Approved List of Bacterial Names (Skerman *et al.* 1980). Many of the species of plant pathogenic pseudomonads and xanthomonads were grouped together as *Pseudomonas syringae* or as *Xanthomonas campestris* in the eighth edition of *Bergey's Manual of Determinative Bacteriology* (Doudoroff & Palleroni 1974; Dye & Lelliott 1974); for these bacteria the nomenclature proposed by Young *et al.* (1978) is recommended.

A. Soft rots

The most important of the bacteria causing soft rots of vegetables and fruits (Table 1) are *Erwinia carotovora* (subsp. *carotovora* and subsp. *atroseptica*) and pectolytic strains of *Pseudomonas fluorescens* (*Ps. marginalis* and

TABLE 1
Bacteria which cause soft rots of vegetables and fruits

Bacterium	Produce affected	Reference
Erwinia carotovora subsp. *carotovora*	Most vegetables and some fruit	*
Erwinia carotovora subsp. *atroseptica*	Most vegetables, particularly potatoes	*
Pectolytic biotypes of *Pseudomonas fluorescens* (*Ps. marginalis* and related bacteria)	Many vegetables	Brocklehurst & Lund (1981)
Bacillus polymyxa and other pectolytic *Bacillus* spp.	Potatoes	Dowson (1943, 1944) Jackson & Henry (1946)
	Peppers	Volcani & Dowson (1948)
Pectolytic *Clostridium* spp.	Potatoes	Lund (1979) Lund *et al.* (1981)
E. chrysanthemi	Pineapple	Lim & Lowings (1979)
Pseudomonas alliicola	Onion	Roberts (1973)
Pseudomonas cepacia	Onion	Smith *et al.* (1966)

*Ramsey *et al.* (1959), Ramsey & Smith (1961), Smith *et al.* (1966), McColloch *et al.* (1968), Smith & Wilson (1978).

related organisms, subsequently referred to collectively as *Ps. marginalis*). These bacteria can attack a wide range of types of plant tissue and cause both field disease (Chupp & Sherf 1960) and post-harvest spoilage. The soft-rot organism *E. chrysanthemi* is a plant pathogen which attacks a wide range of tropical and subtropical crops as well as greenhouse plants grown in temperate regions (Pérombelon & Kelman 1980). It has recently been reported to cause decay of seed potatoes in soil at 30°C (Cother 1979). Both *E. carotovora* subsp. *carotovora* and *E. chrysanthemi* were reported as the cause of core rot found in carrots at harvest in Texas (Towner & Beraha 1976). *Erwinia chrysanthemi* also causes fruit collapse of pineapple (Lim & Lowings 1979); much of the wastage occurs in the field but it can also develop after harvest (Thompson 1937). Pectolytic *Bacillus* spp. particularly *B. polymyxa* have been reported as the cause of soft rots of potatoes, particularly at relatively high temperatures (>30°C) (Dowson 1943, 1944; Jackson & Henry 1946), and green peppers (Volcani & Dowson 1948) and cause spoilage of canned fruit and vegetables.

Pectolytic strains of *Clostridium* are frequently implicated in soft rots of potatoes (Kelman 1979; Lund 1979). Pectate-degrading strains of *Flavobacterium* (Lund 1969) are often present on plant surfaces and associated with soft rots, but there appears to be no evidence that they are a primary cause of spoilage.

B. Other defects

Many bacteria cause field diseases, other than soft rots, which affect the market quality of vegetables and fruits and can result in wastage (Table 2). The disease may be evident as an infection of the vascular system or may result in distinct lesions on leaves, fruit, tubers or roots. Much of the affected produce may be rejected during inspection and packing, but inconspicuous lesions may progress during transit and storage, and may allow soft-rot bacteria to enter and cause more extensive spoilage.

TABLE 2
Bacteria which cause field disease other than soft rots affecting the market quality of vegetables and fruit

Bacterium	Produce affected	References
Corynebacterium flaccumfaciens	Bean	Smith *et al.* (1966)
Corynebacterium michiganense	Tomato	McColloch *et al.* (1968)
Corynebacterium sepedonicum	Potato	Smith & Wilson (1978)
Erwinia ananas	Pineapple	Smoot *et al.* (1971)
Pseudomonas ananas	Pineapple	Smoot *et al.* (1971)
Pseudomonas apii	Celery	Smith *et al.* (1966)
Pseudomonas cichorii	Cabbage, lettuce	Ramsey & Smith (1961) Grogan *et al.* (1977)
Pseudomonas lachrymans	Cucumber, honeydew melon	Ramsey & Smith (1961)
Pseudomonas maculicola	Cauliflower	Ramsey & Smith (1961)
Pseudomonas phaseolicola	Bean	Smith *et al.* (1966)
Pseudomonas pisi	Pea	Smith *et al.* (1966)
Pseudomonas solanacearum	Potato	Smith & Wilson (1978)
Pseudomonas syringae	Bean, citrus fruits, (particularly lemon)	Smith *et al.* (1966) Smoot *et al.* (1971)
Pseudomonas tolaasii	Mushroom	Wong & Preece (1979)
Pseudomonas tomato	Tomato	McColloch *et al.* (1968)
Streptomyces scabies	Potato, beet	Smith & Wilson (1978) Ramsey *et al.* (1959)
Xanthomonas campestris	Cabbage, cauliflower	Ramsey & Smith (1961)
Xanthomonas phaseoli	Bean	Smith *et al.* (1966)
Xanthomonas pruni	Peach, nectarine, apricot, plum	Harvey *et al.* (1972)
Xanthomonas vesicatoria	Tomato, pepper, radish	McColloch *et al.* (1968) Segall & Smoot (1962)

Several reports indicate that healthy vegetables, in particular potato, tomato and cucumber, may contain low numbers of saprophytic or in some cases plant-pathogenic bacteria (Hayward 1974). While it is difficult to discount the possibility of contamination with surface bacteria during sampling, some workers have taken very careful precautions to detect and prevent this. Agronomic factors, cultural practices and differences between cultivars may influence the presence of an internal microflora. In some cases there is evidence that defects of produce can be caused by several species of bacteria which form part of the internal bacterial flora.

Evidence from application of *Serratia* to the sepals of young tomato fruits suggested that bacteria from the sepals could become established in the stem depression and from there penetrate into the growing fruit (Samish & Etinger-Tulczynska 1963). In healthy tissue multiplication of endophytic bacteria may be very limited, but subjecting the plant tissue to stress, e.g. by chilling mature-green tomatoes at 4°C for 7 days, can result in multiplication of these bacteria. There is evidence that graywall of tomato (bacterial necrosis) can be incited by non-pectolytic strains of *Bacillus*, *Erwinia* or *Enterobacter* found in healthy fruit and that the incidence of this defect is increased by chilling the fruit (Beraha & Smith 1964; Segall 1967; Stall & Hall 1969). There are also reports that rind necrosis of cantaloup and of watermelon can be incited by several species of bacteria found in the normal internal microflora of these fruits (Thomas 1976; Hopkins & Elmstrom 1977).

The characteristic symptom of bacterial pink disease of pineapple fruit is a dark brown discoloration which develops in the flesh when heated during the canning process. The flesh of infected but unheated fruits may show no discoloration or only a slight pink or brown colour. Each of three species of bacteria have been reported to cause this defect: *Gluconobacter suboxydans*, *Acetobacter aceti* and *Erwinia herbicola* (Cho *et al.* 1978). The bacteria probably enter the fruit when moisture stress prior to flowering is followed by rainfall during the flowering cycle; this sequence gives rise to growth cracks in the base of blossom cups, allowing invasion of the ovaries by the bacteria. It has been suggested that the discoloration of the flesh produced by heating is due to formation by the bacteria of 2,5-diketogluconic acid.

3. The Mechanism of Maceration of Plant Tissue by Soft-rot Bacteria

The integrity of plant tissue and the cohesion between plant cells depends largely on the pectic substances which form a high proportion of the middle lamella and are also an important component of the primary cell wall. There is little information regarding the detailed structure of the middle lamella, whereas that of primary cell walls of dicotyledonous plants has been studied extensively. Pectic substances are polymers consisting of a main chain of α-1, 4-linked D-galactopyranosyluronic acid units interspersed with 1,2-linked rhamnopyranose units. A proportion of the galacturonan units may be methyl-esterified, in which case the polymer is termed a pectin or pectinic acid (Chesson 1980), the de-esterified form of the polymer is known as pectic acid or polygalacturonic acid; the uronide groups may also be acetylated at C_2 and C_3. Side chains composed of neutral sugars, in particular of α-1,3- and α-1,5-linked L-arabinofuranose and β-1,4-linked D-galactopyranose are covalently linked to the rhamnogalacturonan chain, giving a structure which forms a network of cross-links between protein and, via xyloglucan, between cellulose fibrils in the primary cell wall (Albersheim 1975, 1976; Bateman 1976; Bateman & Basham 1976; Chesson 1980).

The ability of soft rot bacteria to attack the middle lamella and primary cell wall causing disintegration of plant tissue results primarily from their ability to form extracellular, endopectic enzymes which attack the galacturonan component in a random manner and catalyse depolymerization by hydrolysis (hydrolases) or by β-elimination (lyases) (Bateman & Basham 1976; Chesson 1980; Rombouts & Pilnik 1980). These enzymes also render the non-uronide components of the cell wall more susceptible to other degradative enzymes. Pectic enzymes which cause tissue maceration simultaneously cause cell death. The enzymes probably act simultaneously on the middle lamella and primary cell wall, solubilizing the former and loosening the structure of the primary wall in such a way that the cell wall is no longer able to support the cytoplasmic membrane under conditions of osmotic stress, and unless the cells are plasmolysed by immersion in a hypertonic medium, cell death occurs due to membrane damage (Bateman 1976; Bateman & Basham 1976). The available evidence suggests that damage to the membrane results from an indirect, rather than a direct effect of the pectic enzymes.

Strains of *E. carotovora* growing in culture media are capable of forming pectinesterase, EC 3.1.1.11 (Anon. 1979*a*), and several enzymes which attack components of pectic substances (Rombouts 1972; Lund 1979). The major extracellular depolymerizing enzymes are an endopectate lyase, EC 4.2.2.2, with an optimum pH of 8·5 and a requirement for calcium ions (Moran *et al.* 1968) and an endopolygalacturonase, EC 3.2.1.15, with an optimum pH of 5·2–5·4 (Nasuno & Starr 1966*a*); other intracellular pectic enzymes may also be formed (Stack *et al.* 1980). The range of pectic enzymes formed by bacteria growing in plant tissue may differ from that formed in culture media (Bateman & Basham 1976; Pupillo *et al.* 1976). In studies involving growth of *E. carotovora* in plant tissue the major depolymerizing pectic enzyme reported has usually been endopectate lyase (Turner & Bateman 1968; Mount *et al.* 1970; Hall 1971; Stephens & Wood 1975), but in soft rots of potato caused by injection of *E. carotovora* subsp. *atroseptica* the pH of the rotted tissue was *ca.* 6·2 and significant amounts of both polygalacturonase and pectate lyase were present (Brocklehurst & Lund unpublished). Purified endopectate lyase enzymes of *E. carotovora* (Mount *et al.* 1970; Stephens & Wood 1975) and of *E. chrysanthemi* (Garibaldi & Bateman 1971) caused electrolyte loss, maceration and cell death of potato tissue. Although studies by Beraha *et al.* (1974) with mutant strains of *E. carotovora* suggested that polygalacturonase was the significant macerating enzyme, more recent genetic experiments with *E. chrysanthemi* indicated that the pectate lyase was the enzyme of primary importance. Donor strains were obtained able to transfer genes including pat^+ (the gene coding for production of pectate lyase); pat^+ but not pat^- recombinants were capable of macerating plant tissue even though both types formed polygalacturonase (Chatterjee & Starr 1977).

Two strains of *Ps. marginalis* grown in culture media formed small quantities of pectinesterase and both strains formed a pectate lyase with an optimum pH of about 8·5 and a requirement for calcium ions; significant amounts of polygalacturonase, assayed at pH 5·2, were formed by one strain, but only trace amounts by the other (Nasuno & Starr 1966*b*). There appears to be little further information on the significance of the polygalacturonase, but there are several reports of the formation of endopectate lyase by *Ps. marginalis* in culture media and in infected tissue (references cited in Rombouts *et al.* 1978). The purified endopectate lyase formed by a strain of *Ps. fluorescens* grown in a culture medium showed maximum activity at pH 9·4 and required calcium ions (Rombouts *et al.* 1978). The organism attacked potato tissue, and pectate lyase but no polygalacturonase was detected in the resulting rots. The purified pectate lyase macerated potato tissue most rapidly at pH 8·0, well below the optimum pH value for initial attack on polygalacturonate; this was attributed to the greater stability of the enzyme below the pH for optimum activity.

In culture medium *B. polymyxa* formed four endopectate lyase enzymes with pH optima between 8·3 and 9·6, but no polygalacturonase (Nagel & Wilson 1970). Strains of a pigmented, soft rot *Clostridium* grown in potato tissue formed pectinesterase and endopectate lyase, but polygalacturonase could not be detected; in contrast, strains of the flax-retting organism *C. felsineum* formed both polygalacturonase and pectate lyase (Lund & Brocklehurst 1978).

In the case of *Ps. cepacia*, a non-fluorescent pseudomonad, strains of which cause 'sour skin' of onions, a rot affecting certain outer, fleshy scales (Burkholder 1950; Gonzalez & Vidaver 1979), a polygalacturonase was reported to be implicated. Inoculation of the bacterium onto onion slices resulted in maceration and a decrease in the pH of the juice from 5·5 to about 4·0. Pectinesterase and polygalacturonase were present in the macerate but pectate lyase was barely detectable, although high levels were formed during growth of the bacterium at pH 7·0 with pectate as substrate. The purified polygalacturonase showed maximum activity at pH 4·4–4·6, it was probably an endoenzyme and caused maceration of onion tissue (Ulrich 1975).

In order to understand the effect of soft-rot bacteria on vegetables and fruit further information is required on the role of the endopolygalacturonase enzymes formed by strains of *E. carotovora* and *Ps. marginalis*, and on the possibility that these enzymes contribute to maceration, particularly when the bacteria attack plant tissue with a low pH.

Following the action of pectic enzymes, other cell wall polymers become accessible to enzymic degradation, while disruption of protoplasts allows release of nutrients which stimulate growth of invading micro-organisms. *Erwinia carotovora* probably does not attack cellulose, but some strains of *E. chrysanthemi* are well known to do so (Dye 1969; Garibaldi & Bateman 1973; Lund 1979). There is little further information regarding the ability of soft-rot bacteria which attack fresh vegetables or fruit to degrade cell wall polymers other than the rhamnogalacturonan; however, during processing of California ripe olives, spoilage involving softening and sloughing of tissue was attributed to both pectolytic and cellulolytic activities of bacteria (Vaughn *et al.* 1969; Patel & Vaughn 1973).

In addition to the soft-rot bacteria and other plant pathogens, numerous other types of pectolytic bacteria have been described; these do not usually cause post-harvest spoilage, but contribute to breakdown of plant tissue in many environments (Rombouts & Pilnik 1980).

4. Factors Affecting Post-harvest Spoilage by Bacteria

The occurrence and extent of post-harvest spoilage is likely to be affected by the incidence and properties of spoilage bacteria, properties of vegetables and fruits and freedom from damage, and environmental conditions.

A. The incidence and properties of spoilage bacteria

Erwinia carotovora is not usually found as a high proportion of the bacteria on the surface of undamaged, aerial parts of vegetables in the UK. Low numbers, below the limit of detection, are probably often present and enrichment can occur when environmental conditions favour spoilage. Many workers have reported that following introduction into soil the numbers of *E. carotovora* decline rapidly to undetectable levels, but they can remain viable for long periods in plant debris. The organism survives on stored potato tubers, and the seed potato is a major means of transmission to the subsequent crop; it is harboured by volunteer potato plants and possibly in the rhizosphere of other plants, e.g. those of cruciferous crops. Transmission from crop to crop can occur via infected machinery, insects and in some conditions by aerosols (Pérombelon & Kelman 1980).

Fluorescent pseudomonads often form a high proportion of the bacteria present on the aerial parts of vegetables and soft-rot strains are frequently one of the main groups of pectolytic bacteria present (Rombouts 1972). On fresh cauliflower curd the numbers of *Ps. marginalis* were equivalent to *ca.* 16% of the count of viable bacteria on Plate Count agar (Oxoid), whereas the numbers of *E. carotovora* were <1% (Lund 1980). Pectolytic strains of *Bacillus* and *Clostridium* probably survive for long periods as spores in soil.

Soft-rot bacteria should therefore probably be regarded as ubiquitous on vegetables, but the numbers present may vary greatly as a result of agronomic and post-harvest factors. Contamination of vegetables or fruit with high numbers of spoilage bacteria is likely to increase decay, e.g. dissemination of *E. carotovora* during washing of tomatoes has been reported to cause an increase in soft rots (Segall 1967, 1968; Segall *et al.* 1977).

B. Properties of vegetables and fruits and freedom from damage

Maturity of a crop may affect susceptibility to damage and to microbial spoilage. Physical damage in the field or during harvesting and handling is a major factor leading to bacterial spoilage, e.g. soil scarring and mechanical harvesting of field-grown tomatoes in the USA results in losses due to bacterial soft rot (Bartz & Crill 1973) and an increased risk of soft rot of potatoes after mechanical harvesting and handling was reported by Lund & Kelman (1977).

The resistance of plant tissue to infection may be influenced by numerous factors (Wood 1967), some of which are listed below.

(i) *Turgidity of the tissue and nutrient availability*

In general, more turgid tissue is more susceptible to bacterial soft rot (Pérombelon & Kelman 1980). The concentration of available nutrients may affect growth of the bacterium and enzyme production; sugar levels in plant tissue may regulate production of cell wall-degrading enzymes by the bacterium and extracts from plant tissue were shown to have a marked effect on formation of pectic enzymes by *E. carotovora* and *Ps. marginalis* in culture media (Zucker *et al.* 1972; Bateman 1976).

(ii) *The presence of inhibitors of bacterial growth or of the action of pectic enzymes*

Many vegetables, fruits and other plants contain antibacterial compounds (Skinner 1955; Nickell 1959; Al-Delaimy & Ali 1970; Mitscher *et al.* 1972); garlic is a particularly potent source and extracts of garlic have been suggested as a means of controlling certain diseases of plants (Ark & Thompson 1959). Phenolic compounds in plant tissue may inhibit pectic enzymes of the pathogen (Eckert 1978), and proteins associated with the cell walls of dicotyledonous plants are capable of inhibiting fungal polygalacturonase (Albersheim 1976).

Except in the case of compounds responsible for lowering the pH of plant tissue, there is little evidence of situations where antibacterial compounds or enzyme inhibitors determine the resistance of vegetables or fruits to bacteria. The importance of the pH of tissue was mentioned earlier. The resistance of ripe tomato fruit to *Xanthomonas vesicatoria* has been attributed to the fact that the pH of such fruit, 4·0–4·6, was low enough to prevent multiplication of the bacterium, whereas green fruit, in which the pH was about 5·0, was attacked (Gardner & Kendrick 1921). However, subsequent work has shown that the pH of tomatoes varies with cultivar and growing location and *increases* with maturity, only occasionally reaching a pH higher than 4·6 (Wolf *et al.* 1979). There have been contradictory reports of the comparative susceptibility of green and red tomatoes to *E. carotovora*, but using low inocula green fruit were reported to be more resistant than pink or red (Bartz & Crill 1978).

In the case of pineapples the pH may vary between 3·8 and 4·5 and it has been claimed that selection of cultivars with a low pH, and cultural practices which result in higher acidity, reduce the incidence of bacterial brown rot (Thompson 1937).

(iii) *The detailed structure of the cell wall, and its vulnerability to attack by pectic enzymes of the pathogen*

Differences between the cell wall structure of different plants may result in differences in susceptibility to pectic enzymes, e.g. red beetroot tissue was much more resistant than potato tissue to pectic enzymes of *E. aroideae* (*E. carotovora* subsp. *carotovora*) and pectin extracted from red beetroot was degraded relatively slowly (Wood 1955). The structure of the cell wall will depend on the age of the plant and on growth conditions; deposition of lignin as tissue matures can mask the polysaccharide components of cell walls, making them resistant to pectic enzymes (Bateman 1976).

(iv) *The ability of the tissue to form morphological or chemical barriers to growth of the bacterium and to activity of pectic enzymes*

This is a particularly important factor in the case of soft rot of potatoes. Processes which probably contribute to the resistance of tubers to *E. carotovora* include: (1) suberization and periderm formation under wounded tissue; (2) oxidation of phenolic compounds to antibacterial quinones; (3) formation of antibacterial compounds such as rishitin in response to infection. All these processes are dependent on oxygen, and the occurrence of extensive soft rots is usually the result of the development of anaerobic conditions within the tuber and consequent inhibition of these reactions (Lund 1979).

C. Environmental conditions of temperature, relative humidity and free water, and gaseous atmosphere

Many vegetables can be stored at 0–1 °C, in order to retard senescence, but cucumbers, tomatoes, peppers and many tropical fruits and vegetables are susceptible to chilling injury when stored at temperatures much below 10°C, and storage at up to 13 °C may be necessary, depending on the duration of storage required (Ryall & Lipton 1972; Eckert 1978). Maincrop potatoes are maintained at 10–15 °C for about two weeks after harvest in order to allow wound healing; they are subsequently stored at temperatures as low as 4°C for ware and canning but if required for other processing, storage temperatures of 7–10°C are used to minimize low temperature sweetening (Burton 1973).

Although strains of *E. carotovora* subsp. *atroseptica* and subsp. *carotovora* will grow at 5 °C, the rate of growth is considerably slower than that of *Ps. marginalis* (Fig. 1) (Brocklehurst & Lund, 1981; Lund & Lau, unpublished) thus bacterial spoilage of vegetables in refrigerated storage at 0–1 °C would be expected to be due to *Ps. marginalis* rather than to *E. carotovora*.

For storage of most vegetables a relative humidity (RH) of about 95% has been recommended (Ryall & Lipton 1972), but RH values of 98–100% can result in decreased loss of water and wilting without incurring an increase in decay provided the temperature is maintainted as low as 0–4°C (Grierson & Wardowski 1978; van den Berg & Lentz, 1978). At high RH values, small fluctuations in temperature (<1 °C) can result in condensation and formation of water films on the surface of stored vegetables, thus leading to an increased risk of bacterial soft rot. An RH of 94–95% is probably low enough to prevent growth of vegetable-spoilage bacteria such as *Ps. marginalis* and *E. carotovora* on fruit or vegetable surfaces.

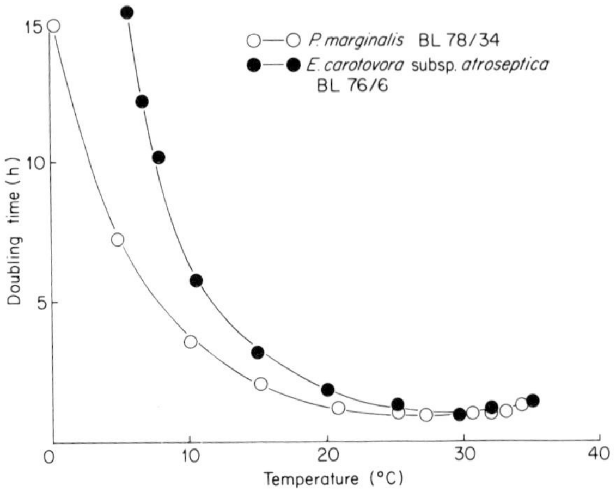

Fig. 1. The effect of temperature on the doubling time of typical strains of *Ps. marginalis* and *E. carotovora* subsp. *atroseptica*. Cultures were aerated by growth in shallow layers of medium with agitation. The medium for *Ps. marginalis* contained (g/l):glucose, 5; vitamin-free Casamino acids (Difco), 1; tryptophane, 0·05; cysteine HCl, 0·05; Yeast extract (Difco), 0·05; plus salts and trace metals; pH 7 (Brocklehurst & Lund 1981). The medium for *E. carotovora* contained (g/l): glucose, 5; Tryptone (Difco), 5; Yeast extract (Difco), 1·5; K_2HPO_4, 0·5; NaCl, 1·5; $MgSO_4.7H_2O$, 0·2; pH7.

Multiplication of these bacteria on vegetables in such conditions will depend on the extent to which a high humidity occurs locally as a result of (a) the humidity of the environment, (b) the local rate of air movement, (c) the presence of cell sap from damaged tissue, (d) the density of packing of produce and (e) the type of packaging. The use of permeable p.v.c. films for wrapping produce during transit and sale can maintain quality by preventing loss of water, but incurs the risk of increased microbial spoilage (Lund 1981) and perforation of these films is advisable unless the packs are maintained at <4°C.

In addition to disseminating spoilage bacteria, washing processes involving immersion in water can cause an increase in infections of potatoes through lenticels and through wounds (Dewey & Barger 1948). Increases in the depth and time of immersion, and the use of water at a lower temperature than the produce have been shown to cause increased bacterial decay of tomatoes and potatoes (Segall *et al.* 1977; Bartz 1981; Kelman & Maher pers. comm.). The presence of a film of water on vegetables in general favours development of soft rot, and on potatoes limits the diffusion of oxygen into tubers resulting in susceptibility to soft rot (Lund 1979).

Controlled or modified atmospheres usually containing oxygen and carbon dioxide concentrations of between 2% and 10%, the precise conditions depending on the commodity, are a possible means, in conjunction with control of temperature, of maintaining the quality of fruits and vegetables during transport and storage (Ryall & Lipton 1972; Eckert 1978; Brecht 1980). The beneficial effect of these atmospheres is probably due mainly to a lowering of the rate of respiration of the produce rather than to an effect on micro-organisms. Reduction of the oxygen concentration in the atmosphere in contact with bacteria to 1–5% is, in general, necessary to reduce the rate of respiration and growth of aerobic bacteria (Brown 1970). In the case of strains of *Ps. fluorescens* at 22°C (Clark & Burki 1972) and *Pseudomonas sp.* at 5°C (Shaw & Nicol 1969) reduction of the oxygen concentration to less than 2% and less than 0·8% respectively was necessary to reduce the rate of growth. In the absence of oxygen *Ps. fluorescens* and *Ps. marginalis* are inhibited, except in the case of denitrifying strains in the presence of nitrate as a hydrogen acceptor, whereas *E. carotovora* is capable of growth at a rate of up to half of the rate in air (Lund & Lau unpublished). According to Wells (1974) carbon dioxide levels higher than 10% were necessary to give any significant inhibition of either *E. carotovora* or *Ps. fluorescens*.

5. Control of Bacterial Spoilage

From the previous discussion it is clear that use of the following measures can contribute to control of bacterial spoilage of vegetables and fruit.
(i) Reduction of infection and dissemination of spoilage bacteria in the field by agronomic measures such as the use of disease-free seed and crop rotation (Crosse 1971; Pérombelon & Kelman 1980).
(ii) Reduction of the numbers of spoilage bacteria in the post-harvest environment, by preventing accumulation of discarded plant debris and by addition of chlorine or chlorine dioxide to water used for washing, hydrocooling and fluming (Eckert 1977).
(iii) Harvest of the crop at the optimum stage of maturity, with the minimum mechanical damage (Eckert 1978).
(iv) Rapid cooling of the produce, e.g. by vacuum cooling or use of an ice-bank cooling system (Derbyshire & Shipway 1978) and storage or transport in controlled, optimum conditions of temperature, relative humidity and gaseous atmosphere in order to maintain the quality of the crop and inhibit multiplication of spoilage micro-organisms (Ryall & Lipton 1972).
(v) Use of antibacterial compounds.

Post-harvest spoilage of vegetables and fruit by bacteria cannot be reliably controlled by the chemicals at present available. Although post-harvest use of medically important antibiotics would not be acceptable, the successful use of streptomycin in the control of certain field diseases caused by bacteria, notably fireblight (Crosse 1971), and the demonstration that treatment with oxytetracycline HCl or with streptomycin sulphate delayed bacterial spoilage of a range of vegetables (Smith 1955; Carroll *et al.* 1957) indicated the possibility of using chemicals to delay or prevent spoilage of plant tissue by bacteria. The fact that chemicals which are both active and acceptable are not yet available may be due largely to the costs of developing and testing new compounds.

Treatment with streptomycin sulphate reduced development of soft rot in potato slices and bacterial decay of seed pieces (Bonde 1953; Bonde & Malcolmson 1956) but increased fungal rotting of seed pieces; further tests have not shown this antibiotic to be effective in preventing soft rot of potatoes (Cates & van Blaricom 1961; Duncan & Gallegly 1963) and its activity may be greatly dependent on the test conditions.

Hypochlorous acid introduced as chlorine gas or as hypochlorite salts at concentrations of a few parts/10^6 of active chlorine are rapidly lethal to bacteria in clear water; concentrations of 50–100 mg/l active chlorine at pH 8·5 are used in commercial practice to reduce the numbers of micro-organisms disseminated in water used to wash, transport and cool vegetables (Eckert 1977). Although immersion in chlorine solutions has been reported to reduce the numbers of food-poisoning and other bacteria on vegetables (Mann 1947; Walters *et al.* 1957) treatment with chlorine has rarely been found to prevent spoilage by micro-organisms already closely associated with plant tissue (Eckert 1977).

In the UK dichlorophen and an iodophor (a nonyl phenoxypoly(ethyleneoxy)ethanol iodine complex) have been cleared under the Pesticides Safety Precautions Scheme for application to certain specified vegetables as fungicides/bactericides (Anon. 1979*b*). Dichlorophen was cleared for use on potatoes, lettuce and tomatoes as a spray or dip containing not more than 0·2% active ingredient (a.i.). The iodophor was cleared for use as a fog, applied by contractors, on apples, white cabbage, carrots, onions and potatoes, or applied as a mist or post-harvest dip to potatoes. These compounds have been claimed to reduce bacterial soft rot of potatoes, but in controlled tests they had no significant effect (Harris 1979; Lund & Wyatt 1979; Wyatt & Lund 1981).

Tests of the ability of other compounds to prevent soft rot of potatoes showed that some, e.g. formaldehyde vapour and sulphur dioxide gas, damaged the tubers at concentrations necessary to reduce soft rot in store. Other compounds including sodium hypochlorite, Anthium Dioxcide (a stabilized form of chlorine dioxide in alkaline solution), freshly generated chlorine dioxide or 8-hydroxyquinoline gave some reduction of soft rot (Wyatt & Lund 1981). Sodium hypochlorite (0·6–2·0 g/l free, available chlorine, pH 9·4–9·7) was the least effective, and in tests incubated for longer times tended to cause increased rotting. At the concentrations of chlorine dioxide required (0·05–0·15%) gas released from solution would probably be toxic to operators and necessitate the use of an enclosed system. Concentrations of Anthium Dioxcide equivalent to 0·05–0·15% of chlorine dioxide were required and the possibility remains that this treatment damages some crops of potatoes. 8-Hydroxyquinoline (0·15–0·3% w/v) gave better control of soft rot than the above compounds (Harris 1979; Wyatt & Lund 1981) but it is doubtful whether this compound would obtain clearance for use on ware potatoes although it may be acceptable for the seed crop.

6. Conclusions

Bacterial spoilage is an important cause of wastage of vegetables and of some fruits. The most damaging type of spoilage is probably that due to soft rot, which results from the action of extracellular pectic enzymes formed by a few types of bacteria, particularly *E. carotovora* and *P. marginalis*, and can be initiated in the field or after harvest. Infection of produce during growth with other bacteria, mainly plant pathogens, can result in defects which reduce the value of the crop and may allow entry of soft-rot bacteria.

The best means of minimizing this wastage are the use of agronomic and post-harvest practices which reduce dissemination of the bacteria, harvesting the crop at the optimum stage of maturity and with the minimum damage, and control of temperature, relative humidity and gaseous atmosphere used for transit and storage. There is a lack of acceptable anti-bacterial compounds to combat bacterial diseases in the field and post-harvest spoilage. In developing countries, where post-harvest losses are already high, some of the changes which may result from improved technology may tend to increase bacterial spoilage. In the developed countries the incidence of bacterial spoilage may tend to be increased by changes in cultivation procedures, e.g. increases in cropping under plastic covers (Zutra *et al.* 1977) in irrigation and in mechanical harvesting, and by changes in marketing and in the pattern of international trade in vegetables and fruit.

I am grateful to Dr R. R. Selvendran of the ARC Food Research Institute, Norwich, for helpful discussions on the structure of plant cell walls, and to Professor A. Kelman of the University of Wisconsin, Madison, USA, for his criticisms of the manuscript.

7. References

ALBERSHEIM, P. 1975 The walls of growing plant cells. *Scientific American* **232**, 81–95.

ALBERSHEIM, P. 1976 The primary cell wall. In *Plant Biochemistry* ed. Bonner, J. & Varner, J. E. London & New York: Academic Press.

AL-DELAIMY, K. S. & ALI, S. H. 1970 Antibacterial action of vegetable extracts on the growth of pathogenic bacteria. *Journal of the Science of Food and Agriculture* **21**, 110–112.

ANON. 1979*a Enzyme Nomenclature* 1978. Report of the Nomenclature Committee, International Union of Biochemistry. London & New York: Academic Press.

ANON. 1979*b United Kingdom Pesticides Safety Precautions Scheme.* Agreed between Government Departments and Industrial Associations. Revised 1979. Recs. 971 (1.12.72) and 1185 (1.5.76). London: Ministry of Agriculture, Fisheries & Food.

ARK, P. A. & THOMPSON, J. P. 1959 Control of certain diseases of plants with antibiotics from garlic (*Allium sativum* L.). *Plant Disease Reporter* **43**, 276–282.

BARTZ, J. A. 1981 The general gas law and post-harvest diseases of tomatoes. *Phytopathology* in press.

BARTZ, J. A. & CRILL, J. P. 1973 A study of methods for reducing bacterial soft rot in wounded fresh market tomatoes. *Proceedings of the Florida State Horticultural Society* **86**, 153–156.

BARTZ, J. A. & CRILL, J. P. 1978 The post-harvest susceptibility of green, pink and red tomato fruit to *Erwinia carotovora* var. *carotovora. Plant Disease Reporter* **62**, 582–585.

BATEMAN, D. F. 1976 Plant cell wall hydrolysis by pathogens. In *Biochemical Aspects of Plant Parasite Relationships* ed. Friend, J. & Threlfall, D. R. London & New York: Academic Press.

BATEMAN, D. F. & BASHAM, H. G. 1976 Degradation of plant cell walls and membranes by microbial enzymes. In *Physiological Plant Pathology* ed. Heitefuss, R. & Williams, P. H. Berlin: Springer-Verlag.

BERAHA, L. & KWOLEK, W. F. 1975 Prevalence and extent of eight market disorders of Western-grown head lettuce during 1973 & 1974 in the Greater Chicago, Illinois area. *Plant Disease Reporter* **59**, 1001–1004.

BERAHA, L. & SMITH, M. A. 1964 A bacterial necrosis of tomatoes. *Plant Disease Reporter* **48**, 558–561.

BERAHA, L., GARBER, E. D. & BILLETER, B. A. 1974 Enzyme profiles and virulence in mutants of *Erwinia carotovora. Phytopathologische Zeitschrift* **81**, 15–22.

BONDE, R. 1953 Preliminary studies on the control of bacterial decay of the potato with antibiotics. *American Potato Journal* **30**, 140–147.

BONDE, R. & MALCOLMSON, J. F. 1956 Studies in the treatment of potato seed pieces with antibiotic substances in relation to bacterial and fungous decay. *Plant Disease Reporter* **40**, 615–619.

BRECHT, P. E. 1980 Use of controlled atmospheres to retard deterioration of produce. *Food Technology* **34**, 45–50.

BROCKLEHURST, T. F. & LUND, B. M. 1981 Properties of pseudomonads causing spoilage of vegetables stored at low temperature. *Journal of Applied Bacteriology* **50**, 259–266.

BROWN, D. E. 1970 Aeration in submerged culture of microorganisms. In *Methods in Microbiology* Vol. 2, ed. Norris, J. R. & Ribbons, D. W. London & New York: Academic Press.

BURKHOLDER, W. H. 1950 Sour skin, a bacterial rot of onion bulbs. *Phytopathology* **40**, 115–117.

BURTON, W. G. 1973 Environmental requirements in store as determined by potential deterioration. *Proceedings 7th British Insecticide and Fungicide Conference* Vol. 3, pp.1037–1055.

CARROLL, V. J., BENEDICT, R. A. & WRENSHALL, C. L. 1957 Delaying vegetable spoilage with antibiotics. *Food Technology* **11**, 490–493.

CATES, F. B. & VAN BLARICOM, L. O. 1961 The effect of several compounds on post-harvest decay of potatoes. *American Potato Journal* **38**, 175–181.

CEPONIS, M. J. 1970 Diseases of California head lettuce on the New York market during the spring and summer months. *Plant Disease Reporter* **54**, 964–966.

CEPONIS, M. J. & BUTTERFIELD, J. E. 1973 The nature and extent of retail and consumer losses in apples, oranges, lettuce, peaches, strawberries and potatoes marketed in greater New York. United States Department of Agriculture Marketing Research Report 996.

CEPONIS, M. J. & BUTTERFIELD, J. E. 1974*a* Causes of cullage of Florida Bell peppers in New York wholesale and retail markets. *Plant Disease Reporter* **58**, 367-369.

CEPONIS, M. J. & BUTTERFIELD, J. E. 1974*b* Market losses in Florida cucumbers and Bell peppers in Metropolitan New York. *Plant Disease Reporter* **58**, 558-560.

CEPONIS, M. J. & BUTTERFIELD, J. E. 1979 Losses in fresh tomatoes at the retail and consumer levels in the Greater New York area. *Journal of the American Society for Horticultural Science* **104**, 751-754.

CEPONIS, M. J., KAUFMAN, J. & BUTTERFIELD, J. E. 1970 Relative importance of gray mold rot and bacterial soft rot of western lettuce on the New York market. *Plant Disease Reporter* **54**, 263-265.

CHATTERJEE, A. K. & STARR, M. P. 1977 Donor strains of the soft-rot bacterium *Erwinia chrysanthemi* and conjugational transfer of the pectolytic capacity. *Journal of Bacteriology* **132**, 862-869.

CHESSON, A. 1980 Maceration in relation to the post-harvest handling and processing of plant material. *Journal of Applied Bacteriology* **48**, 1-45.

CHO, J. J., ROHRBACH, K. G. & HAYWARD, A. C. 1978 An *Erwinia herbicola* strain causing pink disease of pineapple. In *Proceedings 4th International Conference on Phytopathogenic Bacteria* Angers, 1978, pp.433-440.

CHUPP, C. & SHERF, A. F. 1960 *Vegetable Diseases and their Control.* New York: The Ronald Press.

CLARK, D. S. & BURKI, T. 1972 Oxygen requirements of strains of *Pseudomonas* and *Achromobacter. Canadian Journal of Microbiology* **18**, 321-326.

COTHER, E. J. 1979 Bacterial seed tuber decay in irrigated sandy soils of New South Wales. *Potato Research* **23**, 75-84.

COURSEY, D. G. & BOOTH, R. H. 1972 The post-harvest phytopathology of perishable tropical produce. *Review of Plant Pathology* **51**, 751-765.

CROSSE, J. E. 1971 Prospects for the use of bactericides for the control of bacterial diseases. *Proceedings 6th British Insecticide and Fungicide Conference* Vol. 3, pp.694-705.

DERBYSHIRE, D. M. & SHIPWAY, M. R. 1978 Control of post-harvest deterioration in vegetables in the UK. *Outlook on Agriculture* **9**, 246-252.

DEWEY, D. H. & BARGER, W. R. 1948 The occurrence of bacterial soft rot on potatoes resulting from washing in deep vats. *Proceedings of the American Society for Horticultural Science* **52**, 325-330.

DOUDOROFF, M. & PALLERONI, N. J. 1974 *Pseudomonas.* In *Bergey's Manual of Determinative Bacteriology* 8th edn, ed. Buchanan, R. E. & Gibbons, N. E. Baltimore: Williams & Wilkins.

DOWSON, W. J. 1943 Spore-forming bacteria in potatoes. *Nature, London* **152**, 331.

DOWSON, W. J. 1944 Spore-forming bacteria pathogenic to plants. *Nature, London* **154**, 557.

DUNCAN, H. E. & GALLEGLY, M. E. 1963 Field trials for chemical control of seedpiece decay and blackleg of potato. *American Potato Journal* **40**, 279-284.

DYE, D. W. 1969 A taxonomic study of the genus *Erwinia* II The *carotovora* group. *New Zealand Journal of Science* **12**, 81-97.

DYE, D. W. & LELLIOTT, R. A. 1974 *Xanthomonas.* In *Bergey's Manual of Determinative Bacteriology* 8th edn, ed. Buchanan, R. E. & Gibbons, N. E. Baltimore: Williams & Wilkins.

ECKERT, J. W. 1975 Post-harvest diseases of fresh fruits and vegetables — etiology and control. In *Postharvest Biology and Handling of Fruits and Vegetables* ed. Haard, N. F. & Salunke, D. K. Westport, Connecticut: Avi Publishing.

ECKERT, J. W. 1977 Control of postharvest diseases. In *Antifungal Compounds* Vol. 1, ed. Siegel, M. R. & Sisler, H. D. New York: Marcel Dekker.

ECKERT, J. W. 1978 Pathological diseases of fresh fruits and vegetables. In *Postharvest Biology and Biotechnology* ed. Hultin, H. O. & Milner, M. Westport, Connecticut: Food & Nutrition Press.

GARDNER, M. W. & KENDRICK, K. B. 1921 Bacterial spot of tomato. *Journal of Agricultural Research* **21**, 123–156.

GARIBALDI, A. & BATEMAN, D. F. 1971 Pectic enzymes produced by *Erwinia chrysanthemi* and their effects on plant tissue. *Physiological Plant Pathology* **1**, 25–40.

GARIBALDI, A. & BATEMAN, D. F. 1973 Pectolytic, cellulolytic and proteolytic enzymes produced by isolates of *Erwinia chrysanthemi* Burk., McFad et Dim. *Phytopathologia Mediterranea* **12**, 30–35.

GONZALEZ, C. F. & VIDAVER, A. K. 1979 Bacteriocin, plasmid and pectolytic diversity in *Pseudomonas cepacia* of clinical and plant origin. *Journal of General Microbiology* **110**, 161–170.

GRIERSON, W. & WARDOWSKI, W. F. 1978 Relative humidity effects on the post-harvest life of fruits and vegetables. *Hortscience* **13**, 570–574.

GROGAN, R. G., MISAGHI, I. J., KIMBLE, K. A., GREATHEAD, A. S., RIRIE, D. & BARDIN, R. 1977 Varnish spot, destructive disease of lettuce in California caused by *Pseudomonas cichorii*. *Phytopathology* **67**, 957–960.

HALL, J. A. 1971 The maceration of tissues and killing of cells by *Erwinia carotovora* and *Corticium praticola*. PhD Thesis, University of London.

HARRIS, R. I. 1979 Chemical control of bacterial soft rot of wounded potato. *Potato Research* **22**, 245–249.

HARVEY, J. M. 1978 Reduction of losses in fresh market fruits and vegetables. *Annual Review of Phytopathology* **16**, 321–341.

HARVEY, J. M., SMITH, W. L., Jr. & KAUFMAN, J. 1972 Market Diseases of Stone Fruits: Cherries, Peaches, Nectarines, Apricots and Plums. *Agriculture Handbook* No. 414. Washington, DC: United States Department of Agriculture.

HAYWARD, A. C. 1974 Latent infections by bacteria. *Annual Review of Phytopathology* **12**, 87–97.

HOPKINS, D. L. & ELMSTROM, G. W. 1977 Etiology of watermelon rind necrosis. *Phytopathology* **67**, 961–964.

JACKSON, A. W. & HENRY, A. W. 1946 Occurrence of *Bacillus polymyxa* (Praz.) Mig. in Alberta soils with special reference to its pathogenicity in potato tubers. *Canadian Journal of Research* Section C **24**, 39–46.

KELMAN, A. 1979 The role of clostridia in bacterial soft rot of potato. *Report of the Planning Conference on the Developments in the Control of Bacterial Diseases of Potatoes* pp.125–130. Lima, Peru: International Potato Center.

LIM, W. H. & LOWINGS, P. H. 1979 Pineapple fruit collapse in peninsular Malaysia: Symptoms and Varietal Susceptibility. *Plant Disease Reporter* **63**, 170–174.

LUND, B. M. 1969 Properties of some pectolytic, yellow pigmented, Gram-negative bacteria isolated from fresh cauliflowers. *Journal of Applied Bacteriology* **32**, 60–67.

LUND, B. M. 1979 Bacterial soft rot of potatoes. In *Plant Pathogens* ed. Lovelock, D. W. & Davies, R. Society for Applied Bacteriology Technical Series No. 12. London & New York: Academic Press.

LUND, B. M. 1981 The effect of bacteria on post-harvest quality of vegetables. In *Quality in Stored and Processed Vegetables and Fruit* ed. Goodenough, P. W. & Cutting, C. V. 7th Long Ashton Symposium. London & New York: Academic Press.

LUND, B. M. & BROCKLEHURST, T. F. 1978 Pectic enzymes of pigmented strains of *Clostridium*. *Journal of General Microbiology* **104**, 59–66.

LUND, B. M., BROCKLEHURST, T. F. & WYATT, G. M. 1981 *Clostridium puniceum* sp. nov., a pink-pigmented, pectolytic bacterium. *Journal of General Microbiology* **122**, 17–26.

LUND, B. M. & KELMAN, A. 1977 Determination of the potential for development of bacterial soft rot of potatoes. *American Potato Journal* **54**, 211–225.

LUND, B. M. & WYATT, G. M. 1979 A method of testing the effect of antibacterial compounds on bacterial soft rot of potatoes, and results for preparations of dichlorophen and sodium hypochlorite. *Potato Research* **22**, 191–202.

MCCOLLOCH, L. P., COOK, H. T. & WRIGHT, W. R. 1968 *Market Diseases of Tomatoes, Peppers, and Eggplants*. Agriculture Handbook No. 28. Washington, DC: United States Department of Agriculture.

MANN, B. 1947 The control of infected fruit, vegetables and lettuce. *Public Health* **60**, 143–144.

MITSCHER, L. A., LEU, R-P., BATHALA, M. S., WU, W-N. & BEAL, J. L. 1972 Antimicrobial agents from higher plants. 1. Introduction, Rationale and Methodology. *Lloydia* **35**, 157–166.

MORAN, F., NASUNO, S. & STARR, M. P. 1968 Extracellular and intracellular polygalacturonic acid trans-eliminases of *Erwinia carotovora*. *Archives of Biochemistry and Biophysics* **123**, 298–306.

MOUNT, M. S., BATEMAN, D. F. & BASHAM, H. G. 1970 Induction of electrolyte loss, tissue maceration and cellular death of potato tissue by an endopolygalacturonate trans-eliminase. *Phytopathology* **60**, 924–931.

NAGEL, C. W. & WILSON, T. M. 1970 Pectic acid lyases of *Bacillus polymyxa*. *Applied Microbiology* **20**, 374–383.

NASUNO, S. & STARR, M. P. 1966*a* Polygalacturonase of *Erwinia carotovora*. *Journal of Biological Chemistry* **241**, 5298–5306.

NASUNO, S. & STARR, M. P. 1966*b* Pectic enzymes of *Pseudomonas marginalis*. *Phytopathology* **56**, 1414–1415.

NICKELL, L. G. 1959 Antimicrobial activity of vascular plants. *Economic Botany* **13**, 281–318.

PATEL, I. B. & VAUGHN, R. H. 1973 Cellulolytic bacteria associated with spoilage of California Ripe Olives. *Applied Microbiology* **25**, 62–69.

PÉROMBELON, M. C. M. & KELMAN, A. 1980 Ecology of soft rot erwinias. *Annual Review of Phytopathology* **18**, 361–387.

PUPILLO, P. MAZZUCCHI, U. & PIERINI, G. 1976 Pectic lyase isozymes produced by *Erwinia chrysanthemi* Burkh *et al.* in polypectate broth or in *Dieffenbachia* leaves. *Physiological Plant Pathology* **9**, 113–120.

RAMSEY, G. B. & SMITH, M. A. 1961 *Market Diseases of Cabbage, Cauliflower, Turnips, Cucumbers, Melons and Related Crops*. Agriculture Handbook No. 184. Washington DC: United States Department of Agriculture.

RAMSEY, G. B., FRIEDMAN, B. A. & SMITH, M. A. 1959 *Market Diseases of Beets, Chicory, Endive, Escarole, Globe Artichokes, Lettuce, Rhubarb, Spinach and Sweet Potatoes*. Agriculture Handbook No. 155. Washington, DC: United States Department of Agriculture.

ROBERTS, P. 1973 A soft rot of imported onions caused by *Pseudomonas alliicola*. *Plant Pathology* **22**, 98.

ROMBOUTS, F. M. 1972 Occurrence and properties of bacterial pectate lyases. Doctoral Thesis, pp.33–39. Agricultural University, Wageningen, The Netherlands.

ROMBOUTS, F. M. & PILNIK, W. 1980 Pectic enzymes. In *Economic Microbiology* Vol. 5, *Enzymes and Enzymic Conversions* ed. Rose, A. H. London & New York: Academic Press.

ROMBOUTS, F. M., SPAANSEN, C. H., VISSER, J. & PILNIK, W. 1978 Purification and some characteristics of pectate lyase from *Pseudomonas fluorescens* GK-5. *Journal of Food Biochemistry* 2, 1–22.

RYALL, A. L. & LIPTON, W. J. 1972 *Handling, Transportation, and Storage of Fruits and Vegetables*. Westport, Connecticut: AVI Publishing.

SAMISH, Z. & ETINGER-TULCZYNSKA, R. 1963 Distribution of bacteria within the tissue of healthy tomatoes. *Applied Microbiology* **11**, 7–10.

SEGALL, R. H. 1967 Bacterial soft rot, bacterial necrosis and *Alternaria* rot of tomatoes as influenced by field washing and post-harvest chilling. *Plant Disease Reporter* **51**, 151–152.

SEGALL, R. H. 1968 Reducing post-harvest decay of tomatoes by adding a chlorine source and the surfactant Santomerse F85 to water in field washes. *Proceedings of the Florida State Horticultural Society* **81**, 212–214.

SEGALL, R. H. & SMOOT, J. J. 1962 Bacterial black spot of radish. *Phytopathology* **52**, 970–973.

SEGALL, R. H., HENRY, F. E. & DOW, A. T. 1977 Effect of dump-tank water temperature on the incidence of bacterial soft rot of tomatoes. *Proceedings of the Florida State Horticultural Society* **90**, 204–205.

SHAW, M. K. & NICOL, D. J. 1969 Effect of the gaseous environment on the growth on meat of some food poisoning and food spoilage organisms. In *Proceedings 15th European Meeting of Meat Technology*. Helsinki, Finland: University of Helsinki.

SKERMAN, V. B. D., MCGOWAN, V. & SNEATH, P. H. A. 1980 Approved lists of bacterial names. *International Journal of Systematic Bacteriology* **30**, 225–420.

SKINNER, F. A. 1955 Antibiotics. In *Modern Methods of Plant Analysis* ed. Paech, K. & Tracey, M. V. Berlin: Springer-Verlag.

SMITH, M. A., MCCOLLOCH, L. P. & FRIEDMAN, B. A. 1966 *Market Diseases of Asparagus, Onions, Beans, Peas, Carrots, Celery and Related Vegetables*. Agriculture Handbook No. 303. Washington DC: United States Department of Agriculture.

SMITH, W. L., Jr 1955 Streptomycin sulphate for the reduction of bacterial soft rot of packaged spinach. *Phytopathology* **45**, 88–90.

SMITH, W. L., Jr & WILSON, J. B. 1978 *Market Diseases of Potatoes*. Agriculture Handbook No. 479. Washington, DC: United States Department of Agriculture.

SMOOT, J. J., HOUCK, L. G. & JOHNSON, H. B. 1971 *Market Diseases of Citrus and other Subtropical Fruits*. Agriculture Handbook No. 398. Washington DC: United States Department of Agriculture.

STACK, J. P., MOUNT, M. S., BERMAN, P. M. & HUBBARD, J. P. 1980 Pectic enzyme complex from *Erwinia carotovora*: A model for degradation and assimilation of host pectic fractions. *Phytopathology* **70**, 267–272.

STALL, R. E. & HALL, C. B. 1969 Association of bacteria with graywall of tomato. *Phytopathology* **59**, 1650–1653.

STEPHENS, G. J. & WOOD, R. K. S. 1975 Killing of protoplasts by soft rot bacteria. *Physiological Plant Pathology* **5**, 165–181.

THOMAS, C. E. 1976 Bacterial rind necrosis of cantaloup. *Plant Disease Reporter* **60**, 38–40.

THOMPSON, A. 1937 Pineapple fruit rots in Malaya. *Malaysian Agricultural Journal* **30**, 407–420.

TOWNER, D. B. & BERAHA, L. 1976 Core-rot: a bacterial disease of carrots. *Plant Disease Reporter* **60**, 357–359.

TURNER, M. T. & BATEMAN, D. F. 1968 Maceration of plant tissues susceptible and resistant to soft-rot pathogens by enzymes from compatible host-pathogen combinations. *Phytopathology* **58**, 1509–1516.

ULRICH, J. M. 1975 Pectic enzymes of *Pseudomonas cepacia* and penetration of polygalacturonase into cells. *Physiological Plant Pathology* **5**, 37–44.

VAN DEN BERG, L. & LENTZ, C. P. 1978 High humidity storage of vegetables and fruit. *Hortscience* **13**, 565–569.

VAUGHN, R. H., KING, A. D., NAGEL, C. W., NG, H., LEVIN, R. E., MACMILLAN, J. D. & YORK, G. D. 1969 Gram negative bacteria associated with sloughing, a softening of California Ripe Olives. *Journal of Food Science* **34**, 224–227.

VOLCANI, Z. & DOWSON, W. J. 1948 A plant disease caused by a spore-forming bacterium under natural conditions. *Nature, London* **161**, 980.

VON SCHELHORN, M. 1951 Control of microorganisms causing spoilage in fruit and vegetable products. *Advances in Food Research* **3**, 429–482.

WALTERS, A. H., DRIVER, B. M. & BAILEY, C. A. 1957 Bacteriological investigation of hypochlorite disinfection of packed watercress. *The Medical Officer* **November 15**, 277–281.

WELLS, J. M. 1974 Growth of *Erwinia carotovora, E. atroseptica* and *Pseudomonas fluorescens* in low oxygen and high carbon dioxide atmospheres. *Phytopathology* **64**, 1012–1015.

WOLF, I. D., SCHWARTAU, C. M., THOMPSON, D. R., ZOTTOLA, E. A. & DAVIS, D. W. 1979 The pH of 107 varieties of Minnesota-grown tomatoes. *Journal of Food Science* **44**, 1008–1010.

WONG, W. C. & PREECE, T. F. 1979 Identification of *Pseudomonas tolaasi*: the white line in agar and mushroom tissue block rapid pitting tests. *Journal of Applied Bacteriology* **47**, 401–407.

WOOD, R. K. S. 1955 Studies in the physiology of Parasitism XVIII. Pectic enzymes secreted by *Bacterium aroideae*. *Annals of Botany, London* **19**, 1–27.

WOOD, R. K. S. 1967 *Physiological Plant Pathology*. Oxford & Edinburgh: Blackwell Scientific.

WYATT, G. M. & LUND, B. M. 1981 The effect of antibacterial products on bacterial soft rot of potatoes. *Potato Research* in press.

YOUNG, J. M., DYE, D. W., BRADBURY, J. F., PANAGOPOULOS, C. G. & ROBBS, C. F. 1978 A proposed nomenclature and classification for plant pathogenic bacteria. *New Zealand Journal of Agricultural Research* **21**, 153–177.

ZUCKER, M., HANKIN, L. & SANDS, D. 1972 Factors governing pectate lyase synthesis in soft rot and non-soft rot bacteria. *Physiological Plant Pathology* **2**, 59–67.

ZUTRA, D., COHN, R. & VOLCANI, Z. 1977 Recent occurrence of new bacterial diseases on vegetable crops. *Phytoparasitica* **5**, 60.

The Production of Foods and Beverages from Plant Materials by Micro-organisms

J. G. CARR

University of Bristol, Long Ashton Research Station
Long Ashton, Bristol, UK

Contents

1. Introduction

IT IS PERHAPS interesting to speculate how these various foods, prepared as the results of microbial activity, originated. Since most go back to time immemorial it must be concluded that they originated as accidents of storage. We should therefore be grateful to our ancestors for selecting out those foods that are palatable and wholesome. We should, perhaps, be especially grateful to those who became ill or died in the process of selection. Even this supreme sacrifice was not enough since today Indonesians still die from the effects of bongkrekic acid and toxoflavin, two very poisonous substances that can occur in a mould-ripened coconut food called *bongkrek* (van Veen & Mertens 1934 *a,b*; van Veen 1967).

Returning to happier considerations it is my intention to deal with some foods briefly whereas others will be dealt with in more detail. Those which are for more detailed consideration are cocoa, vinegar and brined olives. This symposium deals with plants and bacteria. If I were to keep strictly to this brief there would be little to discuss. However, there are many microbial food preparations where bacteria play a major part and where other micro-organisms are also important. It is some of these that will be discussed.

2. Beverages and Condiments

A. Wines

This term conjures up the image of grape wine, but it should be remembered that wines are not always made from this material (Table 1). One beverage consumed throughout the tropics is palm wine made from the fermented sap of a number of palm species. It is a beverage, very rich in micro-organisms (Okafor 1972, 1975; Faparusi 1973), which is consumed without filtering. Some of the most important organisms found in this sweet, almost milky white beverage are bacteria of the genus *Zymomonas* which play a part in the alcoholic fermentation.

TABLE 1
Beverages and condiments produced by microbial activity

	Organisms essential in preparation			
Product	Yeasts	Lactic acid bacteria	Acetic acid bacteria	Fungi
Wines	+	+	−	−
Ciders	+	+	−	−
Beer	+	−	−	−
Saké	+	+	−	+
Cocoa	+	+	+	−
Coffee*	+	+	−	−
Vinegar	+	−	+	−
Soy sauce	+	+	−	+

*Said to require the presence of *Erwinia dissolvens*, a bacterium of uncertain identity.

The primary organisms in the preparation of wines are the yeasts. It is, however, true to say that no English wine bottler would bottle a wine containing malic acid, a constituent of grape juice. All satisfactory wines must have undergone a malo-lactic fermentation which has the effect of reducing acidity by the decarboxylation of malic acid. This, at the same time, causes the evolution of CO_2, a phenomenon that the bottler would wish to avoid after bottling is complete. Most lactic acid bacteria of wines can carry out a malo-lactic fermentation and no bottled red wine would today be considered complete without it.

Lactic acid bacteria can also bring about this change in ciders, but the occurrence of these organisms in beer is usually regarded as a sign of disorders. There are beers brewed in Belgium called lambic beer and *geuse* which are very highly acidic and it is thought that part of the acidity is caused by lactic acid bacteria.

B. Saké

This drink is usually regarded as typically Japanese and, like many oriental preparations, involves the use of a mould, *Aspergillus oryzae*, which saccharifies the starch of steamed rice. This product is called *koji*, which may then be added to a paste of boiled rice to produce further sugars. These sugars are then fermented with a yeast called *Saccharomyces saké* which is in reality a strain of *Sacc. cerevisiae*. The end product is a pale yellow beverage containing 13–15% (v/v) ethanol plus about 0·3% lactic acid, the latter having been produced by lactic acid bacteria. In this process the lactic acid bacteria play only a minor role.

C. Soy sauce

With the spread of Chinese restaurants throughout Britain, this condiment has become familiar to more and more people. Unlike some condiments that merely add flavour, soy sauce is nutritious since it contains amino acids and related compounds.

It is prepared by mixing equal parts of steamed soy beans and roasted crushed wheat with the mould *Aspergillus oryzae* previously grown for five days on polished rice. This mixture is spread out in shallow trays and incubated at 30°C for 72 h. It is then removed to earthenware vessels of about 50 gallons capacity and to it is added about 20% brine. This is followed by a fermentation by yeasts and lactic acid bacteria that may last for one to three years after which time the liquor is filtered from the residue and then pasteurized. According to Yong & Wood (1974) the yeasts and bacteria present in the fermentation stage belong respectively to the genera *Zygosaccharomyces* and *Pediococcus*.

D. Coffee

The beans or cherries of coffee are fermented only to remove a mulcilaginous layer lying beneath an outer pulpy layer. After mechanical removal of the pulp, the mucilaginous layer is removed by the action of lactic acid bacteria and fermenting yeasts (Pederson & Breed 1946; Agate & Bhat 1966).

E. Cocoa

This is one of the products that will be discussed in more detail. The plant material is the tree, *Theobroma cacao*, which bears fruit and flowers almost continuously, but approximately two main crops appear each year. The fruit consists of yellow or purple pods about 25 cm long and 12·5 cm in diameter. Inside the pods are some 30–40 beans covered with a white mucilage which tastes fruity. Below the mucilage is the pale pink testa of the seed which encloses two bright purple cotyledons known in the chocolate trade as the 'nib'. From this seemingly unpromising material a beverage and a foodstuff can be made. Historically, chocolate for drinking has been known in Europe since the days of Spain's invasion of Central and South America in the sixteenth century. It is in these parts that the plant is indigenous and where it was harvested and used by the Aztecs. Solid chocolate containing cocoa butter is an invention of the last century.

When the beans are fresh they have no chocolate flavour or aroma and it is only by passing them through the fermentation process that these properties begin to develop. Fermentation also gets rid of the mucilage, much of which runs away as a liquid termed 'sweatings'. It is probably the substances in the mucilage which provide nutrients for the developing micro-organisms.

The two most common ways of fermenting beans are either to arrange them in piles enclosed by plantain leaves or to use perforated boxes (Fig. 1). The piles are characteristic of West Africa whereas the boxes are used in various other parts of the world including Malaysia. The beans remain for about six to seven days during which time the piles are turned twice. The boxes, which are constructed in a stepwise manner are usually turned into the one below every 24 h. The effect of this treatment is to promote intense

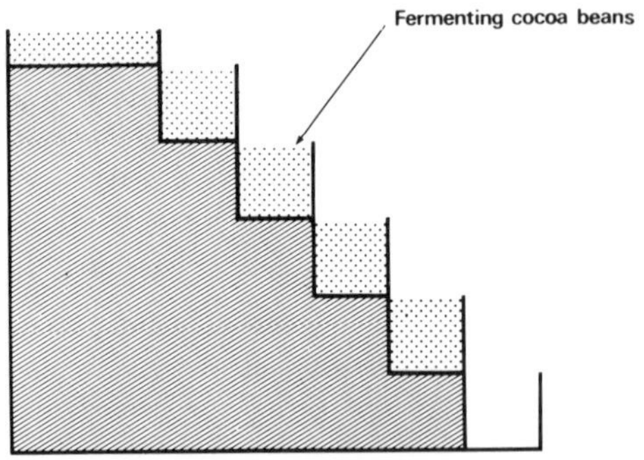

Fig. 1. Side elevation of cocoa fermenting boxes.

microbial activity. The first thing that happens is an alcoholic fermentation of the mucilage which lasts for about 48 h. During this period fermenting yeasts of the genera *Kloeckera*, *Hansenula* and *Saccharomyces* are to be found. As the conditions become less favourable due to increasing alcohol and acetic acid and decreasing amounts of sugar, the main fermenting yeasts disappear. At about this time, but surviving much longer, the lactic acid bacteria appear, the homofermenters being represented by the species *Lactobacillus plantarum* and the heterofermenters by such species as *L. collinoides* (Carr & Davies 1972,1973) and *L. fermentum*. The true acetobacters are the most persistent organisms, being isolatable from the beginning to the end of the fermentation and being represented by such species as *Acetobacter rancens*, *A. lovaniensis*, *A. xylinum*, *A. ascendens*. The other acetic acid bacterium, *Gluconobacter oxydans*, can be isolated in the early stages but it disappears within 24 h.

The appearance of the acids corresponds with the appearance of the organisms that produce them. Thus, acetic acid appears before lactic acid and is the predominant acid, often reaching concentrations of more than 2%. Lactic acid, in contrast, rarely reaches 1%. The source of most of the acetic acid is the ethanol produced by the yeasts: this is converted to the acid by *Acetobacter* spp. Lactic acid is produced mainly as the result of the conversion of available sugars by lactic acid bacteria. However, these organisms can also attack available organic acids, such as citric and malic acid, and convert them mainly to lactic acid. The result of all this microbial activity is to raise the temperature of the beans to 50°C or more. This kills those fermenting yeasts that survive the lethal effect of acetic acid. The heat also kills the beans. As the fermentation progresses increasing amounts of the mucilage are converted to sweatings. Our investigations have shown that the most important groups of organisms present are the yeasts, lactic and acetic acid bacteria. Others are also present; e.g. *Bacillus* spp. can be found occurring sporadically throughout the fermentation, and there are small numbers of other organisms that persist for a long time on the beans during sun-drying.

Eventually all the mucilage disappears and the beans take on an orange-brown hue. The cotyledons of dead beans harden and change in colour either to a slate grey or dark brown. At the end of the fermentation the beans are removed and dried. In Africa this is usually done in the sun, whereas in Malaysia and other parts of the world mechanical means of drying are used. At the end of drying the nib tastes and smells slightly of chocolate, but is mainly very bitter and astringent; at this stage it is hard to believe that this material will eventually become chocolate. When received at the factory the beans are roasted and it is at this stage that the true chocolate flavour is developed.

The microbial picture is very complex and much more work is required to find out which are the working organisms and what substrates they use. Such studies would go a long way to solving the acidity problem in Malaysia in which the final pH of the cotyledons is an undesirable 4·5 – 4·8 instead of 5·2, as in Africa. Cocoa has been the subject of study in various parts of the world and the following represent some of the work done: Roelfson (1958); De Carmargo *et al.* (1963); Rohan (1964); Rohan & Stewart (1966).

F. Vinegar

The main plant materials from which vinegar is made are malted barley, as used in the UK; apples, which are mainly used in the USA; grapes, as used in France and other wine-making countries. There are, however, many other materials used around the world that include citrus fruits, dates, palm wine, peaches, etc. Indeed, any plant material containing fermentable carbohydrates makes a suitable substrate for vinegar. The first process, therefore, is extraction of the soluble nutrients from the plant materials and the second is fermentation by yeasts to produce ethanol. Sometimes the fermentation is contaminated with lactic acid bacteria, but this is not regarded as too serious as the lactic acid produced acts as an alternative substrate to ethanol.

The organisms responsible for the conversion of ethanol to acetic acid are those named as *Acetobacter* spp. in the eighth edition of *Bergey's Manual of Determinative Microbiology* (De Ley & Frateur 1974). These are Gram negative aerobic rods with peritrichous flagella if motile. They are not only able to convert ethanol to acetic acid as shown in equation (1) but they can also overoxidize the acetic acid according to equation (2) when the ethanol has completely oxidized to acetic acid.

$$C_2H_5OH + O_2 \longrightarrow CH_3COOH + H_2O + \text{Energy} \quad (1)$$

$$CH_3COOH + 2O_2 \longrightarrow 2H_2O + 2CO_2 + \text{Energy} \quad (2)$$

These bacteria are often capable of producing high concentrations of acetic acid. One reason for this is that acetic acid is such a weakly dissociated acid that the pH does not drop markedly below 3·0 irrespective of the concentration. Since most strains of *Acetobacter* are able to function at that pH provided nutrients and substrate are present they will produce high concentrations of acetic acid.

There are three main methods of making vinegar. The first of these is called the 'Orleans process' and consists of keeping a cask about half full of wine (Fig. 2). This is inoculated initially with wine already undergoing acetification and the mixture is left in the cask. Gradually the fresh wine acetifies and the cells added in the inoculum form a pellicle on the surface. Such pellicles vary in strength and if they break and sink acetification ceases. This is overcome by floating rafts on the surface. The same effect can be achieved by the use of the species *Acetobacter xylinum* which forms

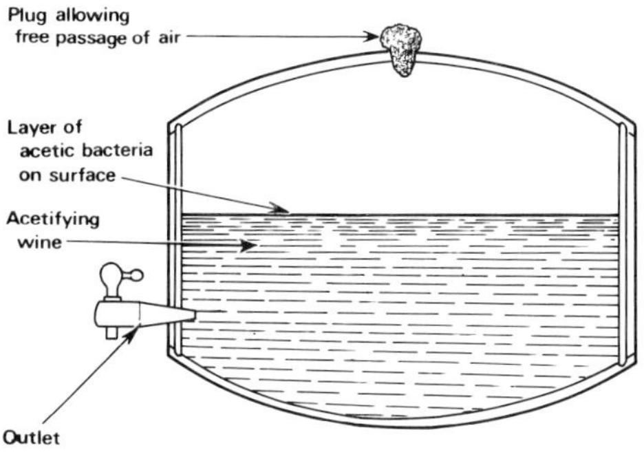

Fig. 2. The Orleans process of vinegar-making.

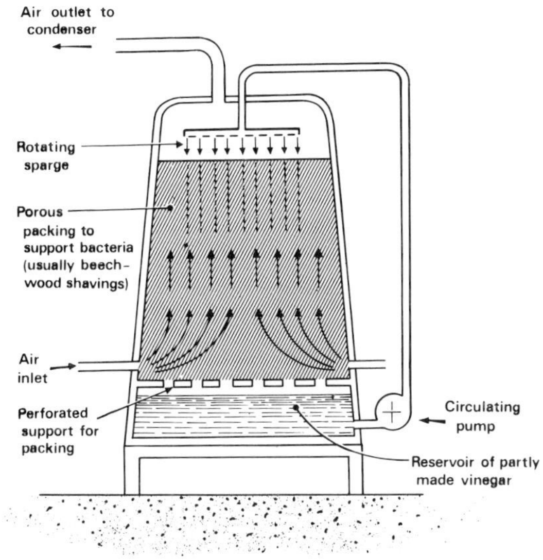

Fig. 3. The quick or trickling vinegar-making process.

a strong buoyant pellicle because it can synthesize threads of cellulose which form a very strong mat.

During acetification a check is kept on the volatile acidity and when this indicates that all the available ethanol has been converted to acetic acid, about half the volume of vinegar is replaced with fresh wine. As a rough guide 1% ethanol gives rise to 1% acetic acid. Although the Orleans process is slow it produces very high quality vinegar.

The next process was intended to replace the Orleans system since it is quicker and is usually called the 'quick' or 'trickling' process (Fig. 3). The plant is usually of wood since vinegar is highly corrosive. It consists of a tower with a perforated false bottom built several feet above the real bottom. Above this is situated a number of holes or vents in the side of the tower. These can be opened or closed to control the rate of air movement into the apparatus. At the top is a vent fitted with a trap to catch volatiles. In the centre of the top is an inlet through which runs a pipe servicing a rotating sparge. The perforated false bottom supports a non-crushable porous material. This varies, but it can be loose twigs, bundles of twigs, wood shavings or washed coke. Its purpose is to provide a non-reacting substratum on which acetic acid bacteria can grow and through which the 'gyle' (vinegar makers' term for fermented beer wort) can trickle over the adhering bacteria. The space under the false bottom acts as a reservoir and to it is connected a pump that conveys the wash to the top of the tower and through the rotating sparge. When the generator is working well the mass of twigs becomes warm due to the exothermic nature of acetification. This causes the tower to act like a chimney and draws cold air in through the ports at the bottom. This in turn influences the temperature of the generator and control is maintained over this by opening or closing the air holes as required.

This process earned the name 'quick' because it was a good deal faster than the Orleans process. It is, however, quite slow, producing about 1% of acetic acid every 24 hours but having the virtue of reliability. As in the Orleans process a certain amount of the previous batch is left in the generator when recharged. One of the few things that can go wrong with a trickling vinegar generator is gradual blockage due to the growth of *A. xylinum*. While extracellular cellulose production is useful when growth occurs on a still liquid, the same production on the support material of a generator is highly undesirable.

A more recent method of vinegar-making is the use of submerged aerated culture conditions. The vessels for this vary somewhat and may be made of wood or acid-resistant lined steel (Fig. 4). Generally they have a means of introducing large volumes of air, a means of cooling and some device for keeping down froth. The substrate, such as fermented beer wort, inoculated with a suitable culture is introduced into the tank and the air turned on, leading to a rapid growth of organisms and a high rate of acetification. The reaction causes a rise in temperature and, to stop the organisms from pasteurizing themselves, a pre-set thermostat is incorporated in the tank and this controls the flow of cold water through the cooling coils. This system has a high acetification rate but suffers from some disadvantages. If the air supply is cut, or even reduced, when the bacteria are acetifying at a high rate most will die instantly and it may take several days before the survivors recover. The other factor is that contaminated air will cause the generator to

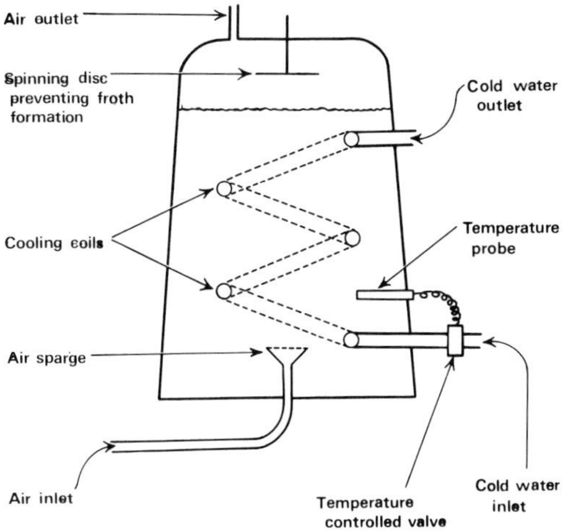

Fig. 4. The submerged vinegar-making process.

cease operating. This is important in cities where contamination of air with SO_2 is common. It is thus essential to maintain a supply of clean air to this kind of vinegar generator.

Vinegars nowadays are stored to improve their quality and are often passed through towers containing beechwood chips to remove potential haze-producing substances. Finally, they are 'filtered bright', pasteurized and bottled.

Unless vinegar is made using micro-organisms it cannot be called 'vinegar' in the UK but must carry the name 'unbrewed condiment'. A further regulation covering vinegar is that it shall contain not less than 4% acid. In general most of the vinegars of high quality contain 5% or more acid. Vinegar-making has been reviewed by Conner & Allgeier (1976) while methods for identifying acetic acid bacteria have been published by Carr & Passmore (1979).

3. Brined and Acidified Products

These are foods which are not particularly popular in Britain but find favour in mainland Europe and particularly in the USA. It is true to say that of all the microbial systems discussed the ones involved in the fermentation of sauerkraut, olives and cucumber involve bacteria alone (Table 2). If these fermentations contain yeasts it usually indicates a disorder. While all of the products previously mentioned undergo fermentation, they are all

TABLE 2
Brined and acidified foods

Product	Organisms essential in preparation			
	Yeasts	Lactic acid bacteria	Acetic acid bacteria	Fungi
Cucumbers	–	+	–	–
Olives	–	+	–	–
Sauerkraut	–	+	–	–
Silage*	–	+	–	–

*Acidified only.

brined and contain fairly high concentrations of salt. There is, however, one acid-fermented product that contains no salt at all. This is silage, in which the low pH inhibits the invasion of less desirable micro-organisms. Although not food for humans it serves as an addition to winter feed for cattle and thus, indirectly, supports our diet.

A. Brined olives

There are two kinds of brined olive, namely green and black. Green olives are harvested before ripening has developed fully. They are first treated with a 'lye' consisting of a 0·9 – 1·25% solution of sodium hydroxide, which is allowed to penetrate the flesh to about half to three-quarters of its depth. This treatment destroys much of the bitter glucoside aleuropein which is present in varying amounts according to the cultivar. The olives are exposed to the lye for about 12 h and this is followed by three or four washes in fresh water to remove the sodium hydroxide. However, when brining is started, the fruit is still quite alkaline. The level of salt used in the brine is initially about 10% and with it is added 3% of lactic acid to neutralize residual alkalinity. The olives remain in this brine for about a week during which time the salinity drops and is finally adjusted to 7·8% salt, which is maintained throughout the fermentation (Vaughn 1975).

Olives are sometimes fermented in casks but more recently large plastic containers have been used. They are sealed over the top with a plastic sheet weighted down with brine (Fig. 5). This not only makes an effective seal but also acts as a valve, allowing fermentation gas out but preventing the entry of air. According to Vaughn (1975) three phases of fermentation can be identified. There is the primary stage in which organisms from the genera *Streptococcus*, *Leuconostoc* and *Pediococcus* occur. At this time there are very few lactobacilli present. If the fermentation does not start properly then organisms of the genera *Enterobacter*, *Escherichia* and similar groups

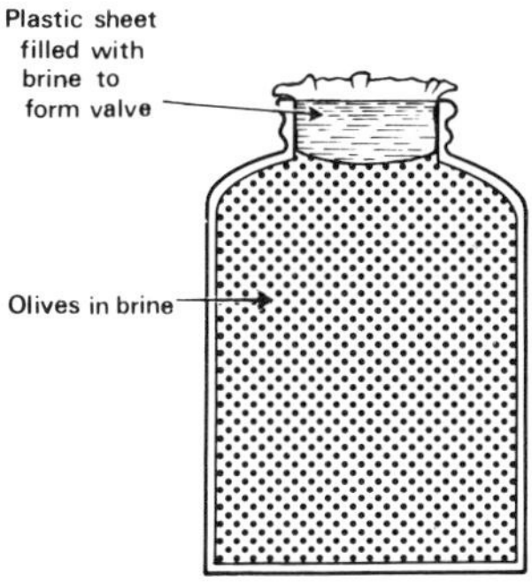

Fig. 5. A large bottle-shaped tank as used in the fermentation of olives.

may be found (Vaughn *et al.* 1969 *a,b*). Yeasts may appear at this time and, if this happens, they can be the cause of sick fermentations (Vaughn *et al.* 1943).

If the fermentation proceeds normally the second or intermediate phase is dominated by lactic acid bacteria. Species of *Leuconostoc* predominate, but there are also *Lactobacillus* spp. which are more acid tolerant than the leuconostocs. All Gram negative bacteria disappear within 12 - 15 days. During three weeks the leuconostocs peak and decline and are then replaced by lactobacilli which appear in large numbers after 21 - 28 days. It is these organisms that bring the pH down to about 4·0 or even lower. If this acidification does not take place a foul-smelling disorder called 'Zapatera spoilage' may develop. This is associated with the presence of *Propionibacterium* and *Clostridium* spp. (Kawatomari & Vaughn 1956; Plastourgos & Vaughn 1957). It is interesting to note that the organism dominating the last stages of fermentation, namely *Lactobacillus plantarum*, can cause a disorder called yeast spot in which various pores on the surface of the olive become the repository of bacterial colonies giving the fruit a white spotty appearance (Vaughn *et al.* 1953).

According to Binstead *et al.* (1962) black olives are allowed to ripen to a straw colour, i.e. just turning red, before processing. They are subject to treatment in the lye and are then alternately soaked in the sodium hydroxide solution and exposed to air. It is this oxidative treatment that causes the fruit to blacken. Having changed colour they are brined in a manner similar to the green olives.

It is not intended to describe the brining of sauerkraut or cucumbers since the underlying principles and occurrence of microbes is similar, and accounts of processing cucumbers and sauerkraut have been written respectively by Etchells *et al.* (1975) and Stamer (1975).

4. Conclusions

This paper is by no means an exhaustive description of plant materials being transformed into various foods by microbial action. For example, sour dough bread (Kline & Sugihara 1971; Sugihara *et al.* 1971) has not been mentioned.

These fermentations, though often caused by similar organisms, produce a wide variety of foods, for what could be more diverse in flavour than soy sauce, vinegar, chocolate and brined olives? It is perhaps fortunate that on this planet there exist such micro-organisms as acetic and lactic acid bacteria whose activities often prevent the depredations brought about by their less acid-tolerant counterparts.

5. References

AGATE, A. D. & BHAT, J. V. 1966 Role of pectinolytic yeasts in the degradation of mucilage layer of *Coffea robusta* cherries. *Applied Microbiology* **14**, 256–260.

BINSTEAD, R., DEVEY, J. D. & DAKIN, J. C. 1962 *Pickle and Sauce Making* 2nd edn, p.239. London: Food Trade Press.

CARR, J. G. & DAVIES, P. A. 1972 The ecology and classification of strains of *Lactobacillus collinoides* nov. spec.: A bacterium commonly found in fermenting apple juice. *Journal of Applied Bacteriology* **35**, 463–471.

CARR, J. G. & DAVIES, P. A. 1973 Designation of a type strain for *Lactobacillus collinoides. Journal of Applied Bacteriology* **37**, 471–472.

CARR, J. G. & PASSMORE, S. M. 1979 Methods for identifying acetic acid bacteria. In *Identification Methods for Microbiologists* 2nd edn, ed. Skinner, F. A. & Lovelock, D. W. pp.33–47. Society for Applied Bacteriology Technical Series No. 14. London & New York: Academic Press.

CONNER, H. A. & ALLGEIER, R. J. 1976 Vinegar: Its history and development. *Advances in Applied Microbiology* **20**, 81–133.

DE CARMARGO, R., LEME, J. Jr & FILHO, A. M. 1963 General observations on the microflora of fermenting cocoa beans (*Theobroma cacao*) in Bahia (Brazil). *Food Technology* **17**, 1328–1330.

DE LEY, J. & FRATEUR, J. 1974 The Genus *Acetobacter*. In *Bergey's Manual of Determinative Bacteriology* 8th edn, ed. Buchanan, R. E. & Gibbons, N. E. pp.276–278, Baltimore: Williams & Wilkins.

ETCHELLS, J. L., FLEMING, H. P. & BELL, T. A. 1975 Factors influencing the growth of lactic acid bacteria during fermentation of brined cucumbers. In *Lactic Acid Bacteria in Beverages and Food* ed. Carr, J. G., Cutting, C. V. & Whiting, G. C. pp.281–305. Proceedings of the Fourth Long Ashton Symposium, 1973. London & New York: Academic Press.

FAPARUSI, S. L. 1973 Origin of initial microflora of palm wine from oil palm trees *Elaeis guineensis*. *Journal of Applied Bacteriology* **36**, 559–565.

KAWATOMARI, T. & VAUGH, R. H. 1956 *Clostridium* associated with zapatera spoilage of olives. *Food Research* **21**, 481–490.

KLINE, L. & SUGIHARA, T. F. 1971 Micro-organisms of the San Francisco sour dough bread process. II Isolation and characterization of undescribed bacterial species responsible for the souring activity. *Applied Microbiology* **21**, 459–465.

OKAFOR, N. 1972 Palm wine yeasts from parts of Nigeria. *Journal of the Science of Food and Agriculture* **23**, 1399–1407.

OKAFOR, N. 1975 Microbiology of palm wine with particular reference to bacteria. *Journal of Applied Bacteriology* **38**, 81–88.

PEDERSON, C. S. & BREED, R. S. 1946 Fermentation of coffee. *Food Research* **11**, 99–106.

PLASTOURGOS, S. & VAUGHN, R. H. 1957 Species of *Propionibacterium* associated with zapatera spoilage of olives. *Applied Microbiology* **5**, 267–271.

ROELFSON, P. A. 1958 Fermentation, drying, and storage of cacao beans. *Advances in Food Research* **8**, 225–296.

ROHAN, T. A. 1964 The precursors of chocolate aroma: A comparative study of fermented and unfermented cocoa beans. *Journal of Food Science* **29**, 456–459.

ROHAN, T. A. & STEWART, T. 1966 The precursors of chocolate aroma: Changes in the sugars during the roasting of cocoa beans. *Journal of Food Science* **31**, 206–209.

STAMER, J. 1975 Recent developments in the fermentation of sauerkraut. In *Lactic Acid Bacteria in Beverages and Food* ed. Carr, J. G., Cutting, C. V. & Whiting, G. C. pp.267–280. Proceedings of the Fourth Long Ashton Symposium, 1973. London & New York: Academic Press.

SUGIHARA, T. F., KLINE, L. & MILLER, M. W. 1971 Micro-organisms of the San Francisco sour dough bread process. I Yeasts responsible for the leavening action. *Applied Microbiology* **21**, 456–458.

VAN VEEN, A. G. 1967 The bongkrek toxins. In *Biochemistry of Some Foodborne Microbial Toxins* ed. Mateles, R. I. & Wogan, G. N. pp.43–50. Massachusetts: Massachusetts Institute of Technology Press.

VAN VEEN, A. G. & MERTENS, W. K. 1934*a* Die Giftstoffe der sogenannten Bongkrekvergiftungen auf Java. *Recueil des Travaux Chimiques des Pays Bas* **53**, 257–268.

VAN VEEN, A. G. & MERTENS, W. K. 1934*b* Das Toxoflavin der gelbe Giftstoff der Bongkrek. *Recueil des Travaux Chimiques des Pays Bas* **53**, 398–404.

VAUGHN, R. H. 1975 Lactic acid fermentation of olives with special reference to California conditions. In *Lactic Acid Bacteria in Beverages and Food* ed. Carr, J. G., Cutting, C. V. & Whiting, G. C. pp.307–323. Proceedings of the Fourth Long Ashton Symposium, 1973. London & New York: Academic Press.

VAUGHN, R. H., DOUGLAS, H. C. & GILLILAND, J. R. 1943 Production of Spanish-type green olives. California Agricultural Experiment Station Bulletin No. 678, pp.1–82.

VAUGHN, R. H., WON, W. D., SPENCER, F. B., PAPPAGIONIS, D., FODA, I. O. & KRUMPERMAN, P. H. 1953 *Lactobacillus plantarum*, the cause of yeast spots on olives. *Applied Microbiology* **1**, 82–85.

VAUGHN, R. H., JAKUBCZYK, T., MACMILLAN, J. D., HIGGINS, T. E., DAVÉ, B. A. & CRAMPTON, V. M. 1969*a* Some pink yeasts associated with softening of olives. *Applied Microbiology* **18**, 771–775.

VAUGHN, R. H., KING, A. D., NAGEL, C. W., NG, H., LEVIN, R. E., MACMILLAN, J. D. & YORK, G. K. II 1969*b* Gram negative bacteria associated with sloughing, a softening of California ripe olives. *Journal of Food Science* **34**, 224–227.

YONG, F. M. & WOOD, B. J. B. 1974 Microbiology and biochemistry of the soy sauce fermentation. *Advances in Applied Microbiology* **17**, 157–194.

Contamination of Food Plants and Plant Products with Bacteria of Public Health Significance

DIANE ROBERTS, G. N. WATSON AND R. J. GILBERT

Food Hygiene Laboratory, Central Public Health Laboratory, London, UK

Contents

1. Introduction

FOODS of plant origin have been incriminated in outbreaks of enteric fever and cholera and on several occasions in episodes of food poisoning. However they do not feature prominently in the communicable disease statistics produced in various parts of the world. Examples of some incidents and the foods involved are given in Table 1, and almost all these outbreaks have involved raw or dehydrated vegetables or fruits.

Plant products further processed, e.g. by canning, particularly home canning, have frequently been implicated in incidents of botulism.

TABLE 1

*Examples of outbreaks of some food-borne infections and intoxications involving foods of plant origin**

Illness	Number affected	Country	Food involved	Reference
Typhoid	49	USA	Celery	Morse (1899)
	110	England	Watercress	Warry (1903)
	2	USA	Rhubarb	Pixley (1913)
	18	USA	Watercress	Anon. (1913)
	97	England†	Lettuce	Dudley (1928)
	4	England	Watercress	Cruickshank (1947)
	339	Germany	Endive	Harmsen (1954)
	>26	Australia	Desiccated coconut	Wilson & MacKenzie (1955)
Paratyphoid	126	Austria	Endive	Donle (1943)
	14	England	Desiccated coconut	Anderson (1960)
Cholera	256	Israel	'Vegetables'	Cohen *et al.* (1971)
Salmonella food poisoning				
Salm. miami	19	USA	Watermelon	Gayler *et al.* (1955)
Salm. bareilly	6	USA	Watermelon	Gayler *et al.* (1955)
Salm. weltevreden	1	Canada	Black pepper	Laidley *et al.* (1974)
Salm. weltevreden	1	Canada	White pepper	Severs (1974)
Salm. oranienburg	18	USA	Watermelon	Anon. (1979*b*)
Salm. typhimurium	133	USA	Celery	Beecham *et al.* (pers. comm.)
Bacillus cereus food poisoning	>350 (>100 outbreaks)	UK, Netherlands, Australia, Canada, Finland, Japan, USA	Rice	Gilbert & Taylor (1976) Gilbert (1979)
	300	Canada	Green bean salad	Schmitt *et al.* (1976)
	4	USA	Soy, cress and mustard sprouts	Portnoy *et al.* (1976)

*Exclusive of botulism, see text.
†British naval vessel.

2. Implication of Foods of Plant Origin in Incidents of Botulism

As *Clostridium botulinum* spores are widely distributed in soil, food, feed and other environments (Sakaguchi 1979), plants and plant products are likely to be contaminated with this organism. However, as the numbers present are probably only small, any public health hazard arises, not from the consumption of fresh vegetables or fruit but from improperly processed and stored products. Botulism is not a problem in the UK as prior to the 1978 canned salmon incident (Ball *et al.* 1979) there had been no reports of this illness for more than 20 years. The last recorded episode of botulism associated with a food of plant origin was in 1935 when there were two cases following the consumption of vegetarian nut brawn (Gilbert 1974). This record is undoubtedly due to the excellent reputation of the canning industry and to the fact that home preservation of non-acid foods such as meat and vegetables, other than by freezing, is actively discouraged. However, outbreaks do occur in other countries such as Poland (Anusz 1971), the USA, the USSR and France (Gilbert 1974), although in many instances the foods involved are of animal origin. *Cl. botulinum* spores present in a food may survive home processing procedures and germinate, grow and elaborate toxins during subsequent storage, particularly at warm ambient temperatures.

In the USA a considerable amount of home canning of vegetables, fruit, meat and fish is carried out. Of 766 food-borne botulism outbreaks reported in the USA from 1899 to 1977, 548 (72%) were from home processed foods (Anon. 1979*a*). Statistics show that during this period 180 of the 297 outbreaks which could be attributed to a specific food and in the majority of which the toxin type was determined, were attributed to vegetables or fruit. Of these 102 (34% of the total) were associated with string beans and most were caused by *Cl. botulinum* type A with a high fatality rate. Other home canned vegetables and fruits involved have included peppers, potatoes, beans, beets, mushrooms, corn, carrots, olives, celery, huckleberry juice, peaches, figs, fruit preserves, apple butter, blackberries and apple sauce. The largest outbreak reported to date in the USA occurred in 1977 when 59 persons were affected with type B botulism after eating at a Mexican-style restaurant in Michigan. The restaurant had used home-canned jalapeno peppers in the preparation of a hot sauce (Terranova *et al.* 1978).

Commercially canned products such as peppers, mushrooms and vichyssoise (potato) soup have also been implicated in incidents of botulism in the USA. The typical cause was underprocessing which may have resulted from inadequate equipment, improper operating procedures and scheduled processes which were not appropriate for the actual operating conditions being used (Lynt *et al.* 1975).

Outbreaks of botulism involving acid foods, i.e. those with a pH of $\leq 4{\cdot}6$ which would include tomatoes, many fruits and pickles, are rare as *Cl. botulinum* cannot grow at these pH levels. However, there can be a hazard of botulism with such foods if other organisms such as moulds are present which grow in the food and raise the pH above 4·6. The main factors which contribute to the ability of *Cl. botulinum* to grow and elaborate toxin in acid foods include (a) contamination with large numbers of *Cl. botulinum* spores, (b) food composition and storage conditions particularly favourable to *Cl. botulinum* growth and toxin production and (c) metabiosis (Odlaug & Pflug 1978).

Sugii & Sakaguchi (1977) suggest several reasons why vegetables, which are probably less nutritious than animal products for growth and toxin production of *Cl. botulinum*, have greater botulogenic properties than foods of animal origin. These include (a) greater chance of contamination with spores, (b) more favourable storage conditions such as absence of curing salts, higher moisture content and storage for a long time and at higher temperatures and (c) possibly larger molecules of toxin are produced with higher oral toxicity, the production of which may depend on quantities of iron and manganese salts in the food.

Apart from the usual food-borne botulism a newly recognized form of the illness, infant botulism, has been reported in which spores of *Cl. botulinum* germinate and multiply in the gut of infants and produce botulinal toxin *in vivo* (Arnon *et al.* 1977; Arnon 1980). Efforts to identify vehicles by which *Cl. botulinum* spores reach the infants' intestines have shown that there is a significant risk factor associated with the ingestion of honey (Chin *et al.* 1979; Midura *et al.* 1979). Approximately one-third of all patients with infant botulism have been fed honey before onset and the organism has been demonstrated in about 30% of honey specimens associated with cases of infant botulism in California. Other sources such as dust from the environment must also play a part in transmission of *Cl. botulinum* to susceptible infants.

3. Contamination of Foods of Plant Origin

A. Vegetables

There has been a number of reports published on the incidence of organisms pathogenic to man in vegetables, fruits and various ingredients prepared from plant material. The results of some of these surveys, carried out in various countries, are summarized in Table 2. A wide range of salmonella serotypes has been isolated from raw vegetables and in several instances

TABLE 2

Reports of isolations of Salmonella *and* Shigella *from vegetables*

Product	Source	Country	Number examined	Number positive	*Salmonella* or *Shigella* isolated	Reference
Lettuce	Vegetable gardens irrigated with residual water	Spain	80	3	*Salm. edinburgh* (1), *Salm. enteritidis* (1) *Salm. kirkee* (1)	Rebollo (1964)
Fruit and vegetables	Market and hospital supplies	Ceylon	1809	10	*Salm. bareilly* (2), *Salm. enteritidis* (1) *Salm. inverness* (1), *Salm. newport* (1) *Salm. riogrande* (1), *Salm. sandiego* (1) *Salm. typhimurium* (1), *Salm. waycross* (2)	Velaudapillai *et al.* (1969)
			1809	8	*Sh. boydii* (2), *Sh. flexneri* (5) *Sh. sonnei* (1)	
Lettuce	Local retail outlets	Italy	125	82	*Salm. typhi* (31), *Salm. anatum* (28) *Salm. dublin* (7), *Salm. schottmuelleri* (34) *Salm. thompson* (7), *Salm. typhimurium* (52)	Ercolani (1976)
Fennel	Local retail outlets	Italy	89	64	*Salm. typhi* (24), *Salm. anatum* (30) *Salm. dublin* (6), *Salm. schottmuelleri* (21) *Salm. thompson* (5), *Salm. typhimurium* (36)	
Raw mixed vegetables	Flight kitchen	Thailand	21	5	*Salm. brunei* (2), *Salm. derby* (2) *Salm. newport* (1)	Vadhanasin *et al.* (1976)
Vegetables	Homegrown and imported	Netherlands	103	23	*Salm. typhi* (1), *Salm. bredeney* (1) *Salm. california* (3), *Salm. heidelberg* (2) *Salm. infantis* (5), *Salm. java* (4) *Salm. montevideo* (5), *Salm. typhimurium* (2) *Salm. unidentified* (1)	Tamminga *et al.* (1978)
Lettuce, endive and watercress	Market	Uruguay	40	5	*Salm. typhi* (1), *Salm. cholerae-suis* (5), *Salm. enteritidis* (2)	Werner de García *et al.* (1978)

included *Salmonella typhi*. In countries with hot climates, often less than perfect sewage systems and where enteric fevers may be endemic, it is not unexpected to find these organisms in raw plant food especially when irrigation is necessary to maintain growing crops. The isolation of salmonellas including *Salm. typhi* from vegetables on sale in the Netherlands, which has a temperate climate similar to that of the UK, led us to carry out a survey of these products on sale by local greengrocers and supermarkets.

Vegetables, mainly of the salad type to be eaten raw, were purchased over the period July 1979 to May 1980. Samples (25 g) were homogenized after chopping (leafy vegetables, bean shoots, celery, mushrooms, spring onions) or peeling (cucumber, tomatoes, aubergine, courgettes—using the skin only) using a Colworth Stomacher (A. J. Seward, UAC House, Blackfriars Road, London SE1 9UG, UK) with 1% buffered peptone water as diluent. Radishes were rinsed in diluent and the rinsing fluid used for further examination as this product releases inhibitory substances on blending (Tamminga *et al.* 1978). The homogenates or rinsings were examined for colony plate count at 35°C, presence of *Escherichia coli* (most probable number method) and the remainder of the suspension used as pre-enrichment culture for salmonellas. The chopped or peeled raw vegetables were also distributed between various enrichment media for the detection of *Staphylococcus aureus*, *Clostridium perfringens*, *Campylobacter* spp., *Shigella*, *Vibrio* spp. and *Yersinia enterocolitica*. The results obtained from 100 samples comprising 16 types of vegetable are given in Table 3.

Plate counts ranged from $<10^2$ to $>10^9$ organisms/g, the lowest being obtained from aubergine, white cabbage and tomato and the highest from bean shoots, cress, spring onions and watercress. Only 14 samples were shown to contain *E. coli*, the highest level being 460/g in a sample of bean shoots. *Salmonella dublin* was isolated from only one (lettuce) of the 100 samples examined. This serotype is particularly associated with bovines but the organism could have reached the lettuce via a number of routes, e.g. through polluted water, fertilizer, sewage sludge or direct faecal pollution from a human or animal source.

Bacillus cereus and *Cl. perfringens* were isolated from approximately one-third of the samples, 35% and 34% respectively. This was not unexpected as these organisms are commonly isolated from the soil and the environment. *Staph. aureus*, *Campylobacter* spp., *Shigella*, *Vibrio* spp. and *Y. enterocolitica* were not isolated from any of the vegetables examined. The majority of samples, where the origin was known, were home grown although a small proportion were imported from the Netherlands, Spain and the USA.

In comparison with the results of other surveys (Table 2) the incidence of

TABLE 3

Microbiological examination of vegetables: results from the Food Hygiene Laboratory (1979–1980)

Vegetable		Colony plate count per g at 37°C (median)	*E. coli* per g (median)	Number of samples containing*		
				B. cereus	*Cl. perfringens*	*Salmonella*
Aubergine	(5)	$2\cdot8 \times 10^3$	<0·3	1	1	0
Bean shoots	(5)	$1\cdot8 \times 10^8$	43	3	3	0
Cabbage (white)	(5)	$8\cdot5 \times 10^3$	<0·3	0	0	0
Cabbage (red)	(5)	$3\cdot0 \times 10^4$	<0·3	1	1	0
Celery	(10)	$4\cdot0 \times 10^6$	<0·3	5	3	0
Coriander	(5)	$1\cdot5 \times 10^7$	Not done	5	5	0
Courgette	(5)	$2\cdot0 \times 10^7$	2·3	2	2	0
Cress	(5)	$2\cdot0 \times 10^8$	15	2	2	0
Cucumber	(10)	$1\cdot7 \times 10^5$	<0·3	3	1	0
Green pepper	(5)	$2\cdot0 \times 10^4$	<0·3	1	1	0
Lettuce	(10)	$1\cdot3 \times 10^7$	<0·3	4	3	1 *Salm. dublin*
Mushrooms	(5)	$2\cdot0 \times 10^6$	<0·3	2	3	0
Radish	(5)	$2\cdot5 \times 10^6$	<0·3	1	2	0
Spring onions	(5)	$5\cdot0 \times 10^7$	<0·3	4	3	0
Tomatoes	(10)	$2\cdot3 \times 10^3$	<0·3	0	1	0
Watercress	(5)	$1\cdot5 \times 10^7$	<0·3	1	3	0
Total	100			35	34	1

**Staph. aureus*, *Campylobacter* spp. *Shigella* spp., *Vibrio* spp. and *Y. enterocolitica* were not isolated from any of these samples.

salmonellas in our studies was low, only 1% compared with 22·3% for Dutch vegetables and 68·2% for Italian lettuce.

B. Herbs and spices

Many materials of plant origin may be unpalatable alone but are extensively used for flavouring and/or colouring other foods. Although they may only be a minor ingredient of a dish they may be used with little or no processing and could possibly carry significant bacterial loads which may render a food hazardous particularly if added after cooking. The group of plant products referred to as herbs and spices covers a wide range of materials including leaves, stems, fruits and bark. For example, from the fruits of the *Capsicum* family come chilli powder (dried pods) and paprika and cayenne pepper (ripe flesh of the red pepper); cinnamon is a dried tree bark while ginger and turmeric are roots. As spices frequently come from areas of the world where sanitary practices are primitive, there may be potential health problems associated with these products. The high incidence of *B. cereus* food poisoning in Hungary has been attributed to the national liking for highly spiced dishes (Ormay & Novotny 1969).

A number of surveys on the incidence of various organisms of public health significance in herbs and spices have been carried out with variable results. Christensen *et al.* (1967) examined black and red pepper from retail outlets, homes and other catering sources and isolated a range of organisms including *E. coli*, *Staphylococcus* sp. and *Bacillus* sp. Levels of 30–700 *Cl. perfringens*/g were found in turmeric, coriander, mustard and fenugreek by Krishnaswamy *et al.* (1971). Similar and even higher levels of this organism (50–2850/g) were reported by Powers *et al.* (1976*b*) who examined samples of bay leaves, cayenne pepper, chilli powder, cinnamon, garlic powder, mustard powder and oregano collected from military bases. Of 114 samples examined, 17 contained *Cl. perfringens* and only one *Staph. aureus*. Later studies on the same range of samples showed that 5 of 110 samples contained *B. cereus* (Powers *et al.* 1976*a*). However in a survey of 10 different spices sampled at port of entry, Julseth & Deibel (1974) reported that "no organisms posing a public health hazard were found". Kaferstein (1976) examining samples of parsley from retail and manufacturing outlets and market gardens in Germany showed that 30 of 64 samples contained *E. coli*.

Although only a few samples appear to contain pathogenic organisms, the colony counts of many of the spices were very high often exceeding 10^7 organisms/g.

Table 4 summarizes the results obtained in a microbiological survey of herbs and spices carried out by the Food Hygiene Laboratory. The samples

TABLE 4

Microbiological examination of herbs and spices: results from the Food Hygiene Laboratory (1979–1980)

Herb or spice		Colony plate count per g at 37°C (median)	Number of samples containing*			
			E. coli > 10 per g	*B. cereus*	*Cl. perfringens*	*Salmonella*
Chilli	(10)	$6{\cdot}0 \times 10^6$	8	4	9	0
Cinnamon	(10)	$1{\cdot}0 \times 10^5$	1	7	7	0
Curry	(10)	$5{\cdot}0 \times 10^6$	6	5	10	0
Ginger	(10)	$1{\cdot}25 \times 10^7$	6	10	9	1 *Salm. rubislaw*
Nutmeg	(10)	$1{\cdot}5 \times 10^4$	3	3	4	0
Oregano	(10)	$2{\cdot}25 \times 10^4$	1	6	6	0
Paprika	(10)	$3{\cdot}0 \times 10^6$	3	3	9	0
Parsley	(10)	$8{\cdot}7 \times 10^4$	2	5	2	0
Pepper (black)	(10)	$4{\cdot}0 \times 10^7$	9	6	10	0
Pepper (white)	(10)	$6{\cdot}5 \times 10^4$	3	10	9	0
Total	100		42	59	75	1

**Staph. aureus* was not isolated from any of these samples.

were purchased either prepacked or loose, from small shops and supermarkets in various parts of England, the majority coming from the London area. Ten samples of 10 different herbs or spices were examined for colony plate count at 35°C and presence of *E. coli*, *B. cereus*, *Cl. perfringens*, *Staph. aureus* and *Salmonella* spp. Counts ranged from $<10^2$ organisms/g to 3×10^8 organisms/g, the highest being found in black pepper, ginger, chilli and curry powder. Forty-two per cent of the samples contained *E. coli* at levels of >10 organisms/g, 59% contained *B. cereus* and 75% *Cl. perfringens*. Only one isolation of salmonella (*Salm. rubislaw*) was made, from a sample of ginger. These results serve to emphasize the fact that herbs and spices are a source of some food poisoning organisms, notably *B. cereus* and *Cl. perfringens*, and can be a means of adding organisms to other foods.

The isolates of *B. cereus* and *Cl. perfringens* from both the vegetable and herbs and spices surveys were examined for specific serotype with the range of 23 *B. cereus* antisera and 152 *Cl. perfringens* antisera currently in use in the Food Hygiene Laboratory. Table 5 summarizes the results of typing these isolates and compares the predominant types found with those occurring most commonly in incidents of food poisoning. Only 31% and 42% of the isolates of *B. cereus* from vegetables and herbs and spices respectively were typable with the available range of sera compared with 90% of strains from food poisoning incidents. However, the figures for these non-food-poisoning isolates are comparable with those reported by Gilbert & Parry (1977) who found that only 45% of isolates from uncooked rice, 47% from cooked rice, 27% from milk and cream and 37% from cooked meat and poultry were typable. The serotypes most frequently found in our surveys were not the types most commonly implicated in food poisoning. *Bacillus cereus* type 1, the predominant food poisoning type, was not isolated from the vegetables and on only two occasions from the herbs and spices.

The serotyping results for *Cl. perfringens* were similar to those obtained for *B. cereus* although overall a greater proportion of the isolates were typable, particularly from the herbs and spices. This may be due to the fact that the range of *Cl. perfringens* antisera is more extensive than for *B. cereus*. The most common types of *Cl. perfringens* found in the two surveys did not compare with the types most commonly implicated in food poisoning apart from type 3/4 which was isolated from both vegetables and herbs and spices. However this type accounted for less than 2·5% of the isolations of *Cl. perfringens*; altogether 83 different serotypes of this organism were found amongst 265 isolates.

TABLE 5

Serotypes of B. cereus *and* Cl. perfringens *from vegetables, herbs and spices compared with the most common types implicated in food poisoning outbreaks*

Organism	Source	Proportion of isolates typable	Number of different serotypes isolated*	Most common serotypes
B. cereus	Vegetables†	31%	7	11/15**; 20
	Herbs & spices‡	42%	14	10; 14; 17; 20
	Food poisoning outbreaks§	90%		1; 5; 8; 12
Cl. perfringens	Vegetables†	52%	24	3/4; 15; TW19
	Herbs & spices‡	71%	76	3/4; 25/68; TW20/PS19
	Food poisoning outbreaks‖	80%		1; 3/4; 11; 11/13; 29; 41

*Using 23 *B. cereus* and 152 *Cl. perfringens* typing antisera.
†Isolates from survey reported in Table 3.
‡Isolates from survey reported in Table 4.
§Gilbert (1979)
‖Smart *et al.* (1979).
**Strains agglutinating with more than one serum.

C. Dried foods including cereals

Some foods of plant origin which are used in a dried form have been allowed to dry naturally while others have undergone some form of heat treatment to remove moisture. This heat treatment is usually sufficient to eliminate many of the vegetative cells present in the raw materials, but sporing organisms such as *Cl. perfringens* and *B. cereus* will survive. The number of organisms present in dried foods will depend on the number and type initially present in the original raw product and the conditions under which they were stored prior to drying. In the dehydrated form these foods remain safe, unless allowed to cross-contaminate other foods which are not to be further processed, but once re-hydrated there will be opportunity for germination of spores and vegetative growth.

TABLE 6
Isolation of B. cereus *from some dried foods*

Product	Number examined	Number positive	Predominant serotypes*	References
Dried potato	20	8 (40%)	Not determined	Kim & Goepfert (1971)
Raw rice	108	98 (91%)	NT; 12; 15; 17	Gilbert & Parry (1977)
Pulses	18	13 (72%)	NT; 2; 8; 10; 20	Blakey & Priest (1980)
Flour	63	14 (22%)	Not determined	Oulsnam (1980)
Cooked rice	456	74 (17%)	NT; 1; 5; 8	Gilbert & Parry (1977)

*NT = not typable.

Table 6 summarizes the data presented in a small number of studies on the incidence of *B. cereus* in some dried foods. A very high proportion of raw rice samples have been found to contain small numbers of *B. cereus* and rice has been implicated on many occasions in the past decade in outbreaks of *B. cereus* food poisoning (Gilbert 1979). The spores survive the initial cooking of the rice and on subsequent storage at warm ambient temperatures germinate, multiply and produce a toxin which can survive a high degree of re-heating. The emetic toxin has been shown to be stable at 126°C for at least 90 min (Melling & Capel 1978). Blakey & Priest (1980) carried out a study on the occurrence of *B. cereus* in dried foods including pulses and cereals because of renewed interest in the use of these products in cooking with the possibility that they may carry toxigenic strains which could be responsible for food poisoning. Their study showed that a variety of pulses and cereals carry *B. cereus* and during normal soaking procedures prior to cooking and also during any subsequent storage the numbers increase to a level at which toxin production could be significant.

D. Fruits and fruit juices

It is generally accepted that most fruits and fruit juices will have a sufficiently acidic pH to inhibit the growth of food poisoning organisms. Also fruits in general are produced some distance from the ground and are less likely to be contaminated by polluted water used for irrigation or by direct faecal contamination apart from bird and insect droppings. However outbreaks have occurred in which fruits and fruit products have been incriminated. Table 1 shows that there have been several episodes of salmonella food poisoning involving pre-cut melon slices. The inner surfaces became contaminated during slicing and the moist flesh of the fruit supported growth of the salmonellas.

In 1974 an outbreak of *Salm. typhimurium* gastroenteritis occurred in the USA following the consumption of non-sterile apple juice (pH 3·4–3·9) (Anon. 1975). The organism was isolated from 154 of 296 patients, six of 30 employees and from two bottles of apple cider. Although this product was called 'apple cider' it is merely an apple juice and not the same product as cider in the UK which is fermented and thus contains a certain proportion of ethanol (at least 4%). Recent experiments (Goverd *et al.* 1979) have shown that *Salm. typhimurium* can survive in apple juices at pH 3·6 or higher for 30 days and can grow well at pH 3·68. Fermentation of the juice decreases the ability of salmonellas to survive, due partly to the presence of ethanol and partly to changes in the nutritional or physiological conditions brought about by the fermenting yeasts. Goverd and co-workers suggest that the production of apple juice should only be contemplated at pH 4·5 or below and that it should be pasteurized before sale.

E. Coconut

Coconut is a plant material which in the past has been a health hazard but with the introduction of legislation in the country of source to enforce minimum hygiene standards the problems have largely been controlled.

In 1953 an outbreak of typhoid fever and salmonellosis associated with contaminated desiccated coconut occurred in Australia (Wilson & MacKenzie 1955). *Salmonella typhi*, *Salm. paratyphi* B and 12 other salmonella serotypes were isolated from samples of the coconut which was imported from Papua (New Guinea). The UK imports many tons of desiccated coconut annually, mainly from Sri Lanka (formerly Ceylon) and the Philippines, for industrial and domestic use in garnishing and flavouring cakes and sweet dishes. In a survey of this product in 1959–1960 (Galbraith *et al.* 1960) 9% of samples from Ceylon were found to be contaminated with salmonellas including *Salm. paratyphi* B. Anderson

(1960) reported an outbreak of paratyphoid infection in which the first patient, who suffered a severe infection, was very fond of eating raw desiccated coconut. The same phage type of *Salm. paratyphi* B was isolated shortly afterwards from shipments of this product. It was obvious that hygienic control of the conditions of processing the material in the producing country was unsatisfactory. Galbraith *et al.* (1960) postulated that the salmonella contamination was of animal origin while *Salm. paratyphi* B originates, directly or indirectly, from contamination of human origin (Anderson 1960).

The inner surface of the coconut kernel should be sterile but during splitting of the husks, paring the rind from the kernels, splitting the flesh into pieces, washing and grinding into meal there is ample opportunity for contamination to be introduced. Meedeniya (1969) carried out investigations at the 'mills' in Ceylon to attempt to trace the source and sequence of salmonella contamination. One of the sources was animal excreta in the yards where the nuts were stacked. Contamination was passed through the successive stages of preparation for manufacture and into the cutter which became the focal point of contamination within the dry section of the mill. To obtain a salmonella-free product with any degree of certainty the cutter must be regularly and efficiently cleaned. Some of the contaminating organisms could survive the drying process.

Regulations were introduced in Ceylon (Anon. 1961) which made it compulsory for the preparation of kernel and the disintegration, drying and packing to be carried out in completely separate parts of the mill and that there should be no access from one part to another. These steps were apparently successful and desiccated coconut became relatively salmonella-free. Table 7 compares the contamination rate found in 1959–1960 with figures compiled from results from 21 laboratories of the Public Health Laboratory Service over the period 1975–1978. There were reports in 1977 and 1978 that shipments of coconut from the Philippines were contaminated with *Salm. senftenberg* but there was no evidence to indicate that there was an increase in isolations of *Salm. senftenberg* in man in 1977 and outbreaks or cases of coconut associated illness were not reported.

F. Cocoa beans

Although there have been incidents incriminating cocoa powder in salmonella food poisoning the incident which caused some concern to the chocolate industry was the outbreak of *Salm. eastbourne* gastroenteritis which occurred in Canada and the USA in 1973–1974. Between December 1973 and mid February 1974, 95 cases of infection due to *Salm. eastbourne* were reported in seven provinces in Canada (D'Aoust *et al.* 1975) and 80

TABLE 7

Isolation of Salmonella *spp. from imported desiccated coconut*

Country of origin	Number examined	Number positive	Serotypes isolated
1959–1960*			
Ceylon (Sri Lanka)	851	76 (9·0%)	*Salm. angoda* (5), *Salm. bareilly* (15) *Salm. hvittingfoss* (2) *Salm. kotte* (2) *Salm. litchfield* (2), *Salm. newport* (6) *Salm. oslo* (1), *Salm. paratyphi B* (15) *Salm. rubislaw* (1), *Salm. senftenberg* (8) *Salm. solna* (3), *Salm. thompson* (2) *Salm. typhimurium* (2), *Salm. unidentified* (1) *Salm. unidentified* probably *Salm. marylebone* (2) *Salm. vancouver* (1), *Salm. waycross* (6), Probable new type (4)
1975–1978†			
Philippines	3991	41	*Salm. anatum* (8), *Salm. derby* (1), *Salm. senftenberg* (34)
Sri Lanka	370	1	
Friendly Isles	30	1	
Other or not recorded	252	0	
(total 1975–1978)	4643	43 (0·9%)	

*Galbraith *et al.* (1960).
†Gilbert & Roberts (1979).

cases from 23 States in the USA (Craven *et al.* 1975). Investigations indicated that Christmas wrapped chocolate balls and other chocolate products manufactured by a single company were the vehicles of infection. The organism was isolated from patients, chocolate products, roasted cocoa beans, the bean cleaning and roasting room and the inside lip of the tempering tank. It is probable that the source of contamination was a particular lot of cocoa beans from Ghana. An interesting feature of this outbreak was that the dose sufficient to cause illness was low, no more than 1000 organisms.

Although the salmonellas were present on the raw beans, probably from faecal contamination, the temperatures employed during the chocolate process, i.e. roasting at 125°C for 30 min, refining at 65°C for 30 min, and conching at 60°C for 5–24 h, should have been sufficient to eliminate the organisms. However, there were many opportunities for cross-contamination, e.g. the roasted beans were returned to the cleaning room for shelling. If recontamination occurred at this point the organisms could well survive the subsequent heating processes, particularly as high decimal reduction times for some salmonella strains have been demonstrated in chocolate suggesting a protective effect by the chocolate mass against damage by heating (Goepfert & Biggie 1968; Barrile & Cone 1970).

Occasionally samples of cocoa powder have been found to contain salmonellas and at least one outbreak of food poisoning has been traced to this source (Anon. 1973). Presumably the cocoa beans were the original source of contamination. The risk, however, is probably only slight and Busta & Speck (1968) have demonstrated that cocoa powder can exert an antimicrobial effect on salmonellas under certain conditions.

G. Health Foods

Health foods have become increasingly popular and many are sold under the label 'organically grown' and without the addition of preservatives or stabilizers. According to Andrews *et al.* (1979) the organic materials used to replenish the soil consist of any parts of a substance which once had life, whether animal, vegetable or metabolic product of an animal or vegetable. Natural animal manures, crop residues and compost are used as plant fertilizers and regulators of soil temperature and moisture content. This combination of 'organic' growth and lack of preservatives could lead to such foods being marketed with high microbiological counts and possibly containing pathogens. The results of a survey of a range of health foods carried out in the USA by Andrews and his co-workers are summarized in Table 8. Although only a small number of samples contained salmonellas, these organisms were present in a wide range of products. It is of note that

TABLE 8

Isolation of Salmonella *spp. from health foods in the USA**

Product	Number examined	Number positive	Source	Salmonella serotypes
Flours (17 types)	186	4	Soy (3), rye (1)	*Salm. cubana* (3), *Salm. molade* (1)
Grains (5 types)	139	1	Brown rice (1)	*Salm. anatum* (1)
Beans and nutmeats (8 types)	343	0	—	—
Seeds and peas (6 types)	138	5	Sunflower (4), alfalfa (1)	*Salm. montevideo* (4), *Salm. poona* (1)
Dietary supplements	328	4	Soy protein powder (2), yeast (2)	*Salm. binza* (1), *Salm. eimsbuettel* (1), *Salm. havana* (1), *Salm. lille* (1), *Salm. molade* (2), *Salm. schwarzengrund* (1)
Total	1134	14		

*Adapted from Andrews *et al.* (1979).

soy protein was found to contain salmonellas on several occasions. This source of protein is being increasingly incorporated into foods particularly in bulk catering such as school meals. Are we in fact adding more harmful organisms to foods which are already a problem considering the high incidence of salmonellas in, and the outbreaks of food poisoning arising from the consumption of, meat and poultry?

4. Source of Organisms

From the preceding paragraphs it can be seen that a proportion of foods of plant origin may contain organisms which can be harmful to man when consumed. The majority of these organisms will lie on the surface of the plant or its fruit and will only enter the inner tissues if there is trauma resulting in exposure of the inner surfaces. This may be from natural means such as birds, insects, animals or the weather or by artificial means such as cutting during harvesting. The main routes of transmission of organisms of public health significance to plants and plant products are via (a) soil, (b) water, (c) fertilizer, (d) direct faecal contamination and (e) poor hygiene in produce handling.

A. Soil

Clostridium perfringens, *Cl. botulinum* and *B. cereus* occur naturally in the soil and therefore may contaminate plants while they grow, when vegetables or fruits fall to the ground and during harvesting. Crops may also be contaminated by dusts blowing over them. Soils in regions remote from habitation rarely contain faecal contamination but land under cultivation may become exposed to such contamination from wild and domestic animals, applications of manure and poor quality irrigation water. Survival times in soil for many enteric pathogens are summarized by Bryan (1977) and range from 1–2 to 259 days for *Salm. typhi* and other salmonellas. Survival of these organisms will depend on a variety of factors including number and type of organism, type of soil, temperature, rainfall, sunlight, protection provided by foliage and competitive microbial flora.

B. Water

Natural surface waters may be polluted by animal and human sewage and use of such water for irrigation, crop spraying, washing and processing or merely for freshening wilted produce on market stalls can lead to contamination of fruit and vegetable crops. Pathogens introduced into a field by irrigation with waste water would, despite considerable reduction in numbers, survive in soil until harvest under some conditions. They are more likely to contaminate crops if the soil is kept moist. Soil also filters micro-organisms and they often become concentrated near areas where plants grow (Bryan 1977).

C. Fertilizers

Many fertilizers contain ingredients prepared from animal waste products such as bone and blood meals and it is well known that such ingredients frequently contain salmonellas (Anon. 1959; Dawkins & Robertson 1967; Patterson 1972). Thus it is possible that food plants could become contaminated from this source. Animal waste slurries and human night soils are also used as fertilizers and are another rich source of pathogenic organisms particularly in areas where enteric diseases are endemic. The use of chicken manure to fertilize celery beds has been suggested as one possible source of contamination in an outbreak of *Salm. typhimurium* food poisoning which occurred in California in 1978 (H. J. Beecham *et al.* pers. comm.). Geldreich & Bordner (1971) in a review of faecal contamination of fruits and vegetables cite instances where the use of night soil on family gardens and small produce farms have been responsible for infestations with round worms and recurrent infections of amoebiasis, bacillary dysentery, enteric fevers, cholera and hookworm in some parts of the world. However, detection of pathogens on farm produce growing in contact with polluted water and soils treated with animal or human manure is infrequent unless the plant samples are grossly contaminated with sewage or are observed to have faecal particles adhering to them.

D. Direct faecal contamination

Direct faecal contamination of fruits, vegetables and other crops can occur by natural means from insects, birds, wild and domestic animals and humans or by agricultural practices as already described. Insects are only a minor problem although apart from depositing their own faecal material on plants they frequently contact animal waste and may transmit organisms from this source on their hairy exoskeletons.

Snails eat faecal matter and may transmit pathogens to plant food such as salad vegetables. Steiniger (1957) examined snails from gardens manured with faecal matter and fields fertilized with sewage sludge and isolated *Salm. paratyphi* B, *Salm. java* and *Salm. infantis* in a batch of five snails. Of 18 snails found in market vegetables, two yielded *Salm. typhi.*

According to Geldreich & Bordner (1971), the random contamination of garden and orchard produce by direct defaecation of wildlife and farm animals is probably the most significant natural means of disease transmission. Leafy vegetable and root crops can support considerable rodent populations which can transmit organisms to the plants via their faecal droppings. The most frequent source of faecal contamination of maturing fruits and berries are birds. Both rodents and birds may harbour a residual level of salmonellas which can be transmitted to man through faecal contamination of fruits and vegetables.

Farm animals which stray into fields and orchards can also contaminate produce by direct defaecation. Another means of transmission from this source is the use of animal manures as fertilizer as already mentioned. These are a particularly hazardous source of pathogens as it is well known that many farm animals, particularly pigs and poultry, can be excreting salmonellas.

E. Poor hygiene in produce handling

Once a crop is harvested the microbiological condition in which it reaches the consumer is dependent on a number of factors: the time, temperature and atmosphere in which it is stored, the amount of damage it receives during storage, washing, processing, packaging and retail display and the amount of contact with insects, birds, animals and humans.

There are opportunities for transmission of organisms of faecal origin from farm labourers, from those involved in processing and packaging and from market workers, particularly in areas where sanitation is poor. Rinse water used in the field for cleaning produce is often of poor quality. Washing may be little more than a dipping of produce in irrigation ditches or static containers of water used many times over. During transport produce may be freshened by spraying with water of doubtful quality. In markets there may be further trimming, sorting and re-packing and sometimes additional spraying with water to maintain freshness before final dispersal. Crops may be transported in dirty freight cars, cargo ships, boats, trucks and other vehicles which are inadequately or seldom cleaned between shipments and there is often easy access for dust, insects, birds and rodents.

5. Survival on Plant Foods

Although there are many opportunities for pathogenic organisms to contaminate farm produce and the surveys reported in Table 2 do show that salmonellas and shigellas can be found in a wide variety of fruits and

TABLE 9
Survival of Vibrio cholerae, Shigella *spp. and* Salmonella *spp. on foods of plant origin*

Organism	Product	Survival time (days)
V. cholerae	Grains and fruits	1-3
	Spices	1-5
	Fresh vegetables	1-7
	Fresh vegetables (refrigerated)	7-10
Sh. flexneri	Tomato (surface)	2
Sh. sonnei	Apple (skin)	8
	Tomato (tissue)	10
Salm. typhi	Lettuce	11-21
	Vegetables (leaves and stems)	10-31
	Radishes	28-53
Salmonella spp.	Cabbage and gooseberries	>5
	Tomatoes	3-20
	Carrots	>10
	Beet leaves	21
	Soil and potatoes	>40

Adapted from Barua (1970) and Bryan (1977).

vegetables, in general enteric pathogens are infrequently isolated from crops in fields or after harvesting. When they are found it is usually for only a short time after irrigation (Bryan 1977). Various studies have been carried out on the survival of enteric pathogens on a range of foods and the results of those relating to foods of plant origin are summarized in Table 9. These data suggest that there is ample time for crops that become contaminated in the field from polluted irrigation water or direct faecal contact to remain contaminated until harvest. Survival of these organisms will be governed by the amount of sunshine, temperature and character of accompanying organisms.

Melick (1917) reported the survival of *Salm. typhi* in soil for 5–8 weeks. He planted radish seeds, inoculated the soil and found that the radishes were still infected for periods of up to 7 weeks later. Ercolani (1979) considers that a single polluting event early during the growth cycle of lettuce in the field may result in the presence of *Salm. typhi* and an increase in indicator organisms in the harvested produce. The figures obtained in his study of the survival of *Salm. typhi* on lettuce in the field, of up to 75 days in winter and 50 days in summer, are higher than those from studies previously reported (Table 9).

Sporing organisms such as *Clostridium* spp. and *Bacillus* spp. will be more resistant to adverse conditions such as sunlight and desiccation and will undoubtedly survive for long periods of time.

6. Discussion

It is evident that there are many opportunities for foods of plant origin to become contaminated with organisms of public health significance. However the numbers present are likely to be low. It is important to prevent the organisms from multiplying to levels at which they can cause infections or intoxications. Salad vegetables and fruits which are likely to be eaten raw should be washed well before consumption and if stored for future use this should be at temperatures which do not permit bacterial multiplication. Foods which are cooked should be safe unless cooking is inadequate and subsequent storage is at warm ambient temperatures, or there is cross-contamination from other raw foods prepared in the same environment.

The food poisoning statistics for England and Wales (Table 10) do not indicate that vegetables and fruits play a prominent part in this illness.

TABLE 10

Outbreaks of bacterial food poisoning and salmonella infection in England and Wales associated with vegetables (1965–1978)

Year	Number of outbreaks	Food	Organism responsible
1965–1968*	1	Fried potatoes	?
1969–1972†	4	Pease pudding (2), creamed potatoes, canned peas	*Cl. perfringens* (2) *Staph. aureus* (2)
1973–1975‡	0	—	—
1976–1978§	3	Pease pudding (2), canned peas	*Cl. perfringens* (3)

*Vernon (1966, 1967, 1969, 1970).
†Vernon & Tillett (1974).
‡Vernon (1977).
§Hepner (1980).

In the 14-year period from 1965–1978 only eight incidents of food poisoning were attributed to the consumption of vegetables out of a total of 1428 outbreaks where the food vehicle was traced. These episodes were associated with vegetables which had received further processing before consumption.

Evidence suggests that in England and Wales, vegetables, particularly those of the salad type which are eaten raw, are not a major source of organisms which are a hazard to health. However, in countries particularly those with less temperate climates, where irrigation often with polluted water is necessary to maintain an adequate water supply to growing crops, where sanitation is primitive and night soil is used as fertilizer, where animals are allowed to roam amongst crops and where the general standard of hygiene is poor, there is a risk of transmitting disease via foods of plant origin.

Diseases caused by micro-organisms other than bacteria, e.g. viral hepatitis (Hutcheson 1971) and fascioliasis (Facey & Marsden 1960; Ashton *et al.* 1970; Hardman *et al.* 1970) may also be transmitted to man via foods of plant origin. Outbreaks of food poisoning have also been attributed to certain foods plants when the cause was non-microbial but due to toxic components occurring naturally in the plant. Examples of these include solanine poisoning from potatoes (Wilson 1959; McMillan & Thompson 1979) and poisoning from raw red kidney beans (Noah *et al.* 1980).

7. References

ANDERSON, E. S. 1960 The occurrence of Salmonella paratyphi B in desiccated coconut from Ceylon. *Monthly Bulletin of the Ministry of Health and the Public Health Laboratory Service* **19**, 172–175.

ANDREWS, W. H., WILSON, C. R., POELMA, P. L., ROMERO, A. & MISLIVEC, P. B. 1979 Bacteriological survey of sixty health foods. *Applied and Environmental Microbiology* **37**, 559–566.

ANON. 1913 A typhoid outbreak apparently due to polluted watercress. (Report of the Philadelphia Bureau of Health). *Engineering News* **70**, 322. Cited by Bryan, F. L. 1977 in *Journal of Food Protection* **40**, 45–56.

ANON. 1959 Salmonella organisms in animal feeding stuffs (Report). *Monthly Bulletin of the Ministry of Health and the Public Health Laboratory Service* **18**, 26–35.

ANON. 1961 Desiccated Coconut (Manufacture and Export) Regulations 1961. Coconut Products Ordinance.

ANON. 1973 Salmonella surveillance other than *S. typhi* and *S. paratyphi* 1971. (Report of the World Health Organization). *Weekly Epidemiological Record* **48**, 377–381.

ANON. 1975 *Salmonella typhimurium* outbreak traced to a commercial apple cider—New Jersey. (Report of the Center for Disease Control). *Morbidity and Mortality Weekly Report* **24**, 87–88.

ANON. 1979*a Botulism in the United States, 1899–1977.* A Handbook for Epidemiologists, Clinicians and Laboratory Workers, May 1979. Center for Disease Control.

ANON. 1979*b Salmonella oranienburg* gastroenteritis associated with consumption of precut watermelons—Illinois. (Report of the Center for Disease Control). *Morbidity and Mortality Weekly Report* **28**, 522–523.

ANUSZ, Z. 1971 Wstępna analiza epidemiologiczna zatruć pokarmowych wywołanych przez *Clostridium botulinum* w Polsce w latach 1959–1969 (Preliminary epidemiological analysis of food poisonings caused by *Clostridium botulinum* in Poland in the years 1959–1969). *Polskiego Tygodnika Lekarskiego* **26**, 1491–1494.

ARNON, S. S. 1980 Infant botulism. *Annual Review of Medicine* **31**, 541–560.

ARNON, S. S., MIDURA, T. F., CLAY, S. A., WOOD, R. M. & CHIN, J. 1977 Infant botulism: epidemiological, clinical and laboratory aspects. *Journal of the American Medical Association* **237**, 1946–1951.

ASHTON, W. L. G., BOARDMAN, P. L., D'SA, C. J., EVERALL, P. H. & HOUGHTON, A. W. J. 1970 Human fascioliasis in Shropshire. *British Medical Journal* **3**, 500–502.

BALL, A. P., HOPKINSON, R. B., FARRELL, I. D., HUTCHINSON, J. G. P., PAUL, R., WATSON, R. D. S., PAGE, A. J. F., PARKER, R. G. F., EDWARDS, C. W., SNOW, M., SCOTT, D. K., LEONE-GANADO, A., HASTINGS, A., GHOSH, A. C. & GILBERT, R. J. 1979 Human botulism caused by *Clostridium botulinum type* E: The Birmingham outbreak. *Quarterly Journal of Medicine* **48**, 473–491.

BARRILE, J. C. & CONE, J. F. 1970 Effect of added moisture on the heat resistance of *Salmonella anatum* in milk chocolate. *Applied Microbiology* **19**, 177–178.

BARUA, D. 1970 Survival of cholera vibrios in food, water and fomites. In Principles of Practice of Cholera Control. World Health Organization Public Health Paper No. 40, pp.29–31.

BLAKEY, L. J. & PRIEST, F. G. 1980 The occurrence of *Bacillus cereus* in some dried foods including pulses and cereals. *Journal of Applied Bacteriology* **48**, 297–302.

BRYAN, F. L. 1977 Diseases transmitted by foods contaminated by waste water. *Journal of Food Protection* **40**, 45–56.

BUSTA, F. F. & SPECK, M. L. 1968 Antimicrobial effect of cocoa on salmonellae. *Applied Microbiology* **16**, 424–425.

CHIN, J., ARNON, S. S. & MIDURA, T. F. 1979 Food and environmental aspects of infant botulism in California. *Reviews of Infectious Diseases* **1**, 693–696.

CHRISTENSEN, C. M., FANSE, H. A., NELSON, G. H., BATES, F. & MIROCHA, C. J. 1967 Microflora of black and red pepper. *Applied Microbiology* **15**, 622–626.

COHEN, J., SCHWARTZ, T., KLASMER, R., PRIDAU, D., GHALAYINI, H. & DAVIES, A. M. 1971 Epidemiological aspects of cholera el tor outbreak in a non endemic area. *Lancet* **ii**, 86–89.

CRAVEN, P. C., MACKEL, D. C., BAINE, W. B., BARKER, W. H., GANGAROSA, E. J., GOLDFIELD, M., ROSENFELD, H., ALTMAN, R., LACHAPELLE, G., DAVIES, J. W. & SWANSON, R. C. 1975 International outbreak of *Salmonella eastbourne* infection traced to contaminated chocolate. *Lancet* **ii**, 788–793.

CRUICKSHANK, J. C. 1947 Typhoid fever in Devon. The value of phage-typing in a rural area. *Monthly Bulletin of the Ministry of Health and the Public Health Laboratory Service* **6**, 88–96.

D'AOUST, J. Y. D., ARIS, B. J., THISDELE, P., DURANTE, A., BRISSON, N., DRAGON, D., LACHAPELLE, G., JOHNSTON, M. & LAIDLEY, R. 1975 *Salmonella eastbourne* outbreak associated with chocolate. *Canadian Institute of Food Science and Technology Journal* **8**, 181–184.

DAWKINS, H. C. & ROBERTSON, L. 1967 Salmonellas in animal feeding-stuffs. *Monthly Bulletin of the Ministry of Health and the Public Health Laboratory Service* **26**, 215–221.

DONLE, W. 1943 Ueber eine Paratyphus A-Epidemie in Wien (An epidemic of paratyphoid A in Vienna). *Zeitschrift für Hygiene und Infektionskrankheiten, Medizinische Mikrobiologie, Immunologie und Virologie* **124**, 683–703.

DUDLEY, S. F. 1928 Some aspects of the epidemiology of typhoid fever in the Royal Navy. *Proceedings of the Royal Society of Medicine* **21**, 785–800.

ERCOLANI, G. L. 1976 Bacteriological quality assessment of fresh marketed lettuce and fennel. *Applied and Environmental Microbiology* **31**, 847–852.

ERCOLANI, G. L. 1979 Differential survival of *Salmonella typhi*, *Escherichia coli*, and *Enterobacter aerogenes* on lettuce in the field. *Zentralblatt für Bakteriologie, Parasitenkunde, Infektionskrankheiten und Hygiene* Abt. II **134**, 402–411.

FACEY, R. V. & MARSDEN, P. D. 1960 Fascioliasis in man: an outbreak in Hampshire. *British Medical Journal* **2**, 619–625.

GALBRAITH, N. S., HOBBS, B. C., SMITH, M. E. & TOMLINSON, A. J. H. 1960 Salmonellae in desiccated coconut an interim report. *Monthly Bulletin of the Ministry of Health and the Public Health Laboratory Service* **19**, 99–106.

GAYLER, G. E., MACCREADY, R. A., REARDON, J. P. & MCKERNAN, B. F. 1955 An outbreak of salmonellosis traced to watermelon. *Public Health Reports* **70**, 311–313.

GELDREICH, E. E. & BORDNER, R. H. 1971 Fecal contamination of fruits and vegetables during cultivation and processing for market. A review. *Journal of Milk and Food Technology* **34**, 184–195.

GILBERT, R. J. 1974 Staphylococcal food poisoning and botulism. *Postgraduate Medical Journal* **50**, 603–611.

GILBERT, R. J. 1979 *Bacillus cereus* gastroenteritis. In *Food-borne Infections and Intoxications* ed. Riemann, H. & Bryan, F. L. pp.495–518. London & New York: Academic Press.

GILBERT, R. J. & PARRY, J. M. 1977 Serotypes of *Bacillus cereus* from outbreaks of food poisoning and from routine foods. *Journal of Hygiene, Cambridge* **78**, 69–74.

GILBERT, R. J. & ROBERTS, D. 1979 Food poisoning risks associated with foods other than meat and poultry — Outbreaks and surveillance studies. *Health and Hygiene* **3**, 33–40.

GILBERT, R. J. & TAYLOR, A. J. 1976 *Bacillus cereus* food poisoning. In *Microbiology in Agriculture, Fisheries and Food* ed. Skinner, F. A. & Carr, J. G. pp.197–213. Society for Applied Bacteriology Symposium Series No. 4. London & New York: Academic Press.

GOEPFERT, J. M. & BIGGIE, R. A. 1968 Heat resistance of *Salmonella typhimurium* and *Salmonella senftenberg* 775W in milk chocolate. *Applied Microbiology* **16**, 1939–1940.

GOVERD, K. A., BEECH, F. W., HOBBS, R. P. & SHANNON, R. 1979 The occurrence and survival of coliforms and salmonellas in apple juice and cider. *Journal of Applied Bacteriology* **46**, 521–530.

HARDMAN, E. W., JONES, R. L. H. & DAVIES, A. H. 1970 Fascioliasis — a large outbreak. *British Medical Journal* **3**, 502–505.

HARMSEN, H. 1954 Die Stuttgarter und Wiener Typhusepidemien als Beispiele der Gefaht fäkaler Salat und Gemüsekopfdüngung durch Verrieselung und Verregnung. (The Stuttgart and Vienna enteric outbreaks as examples of the danger of irrigating or spraying salads and other vegetables with sewage.) *Stadtehygiene Freeburg/Breisgau* **5**, 54–57. Abstracted in *Bulletin of Hygiene* 1954 **29**, 1036.

HEPNER, E. 1980 Food poisoning and *Salmonella* infections in England and Wales, 1976–78. *Public Health, London* **94**, 337–349.

HUTCHESON, R. H. 1971 Infectious hepatitis — Tennessee. *Morbidity and Mortality Weekly Report* **20**, 357.

JULSETH, R. M. & DEIBEL, R. H. 1974 Microbial profile of selected spices and herbs at import. *Journal of Milk and Food Technology* **37**, 414–419.

KAFERSTEIN, F. K. 1976 The microflora of parsley. *Journal of Milk and Food Technology* **39**, 837–840.

KIM, H. U. & GOEPFERT, J. M. 1971 Occurrence of *Bacillus cereus* in selected dry food products. *Journal of Milk and Food Technology* **34**, 12–14.

KRISHNASWAMY, M. A., PATEL, J. D. & PARTHASARATHY, N. 1971 Enumeration of micro-organisms in spices and spice mixtures. *Journal of Food Science and Technology* **8**, 191–194.

LAIDLEY, R., HANDZEL, S., SEVERS, D. & BUTLER, R. 1974 *Salmonella weltevreden* outbreak associated with contaminated pepper. *Epidemiological Bulletin* **18**, 62.

LYNT, R. K., KAUTTER, D. A. & READ, R. B. 1975 Botulism in commercially canned foods. *Journal of Milk and Food Technology* **38**, 546–550.

MCMILLAN, M. & THOMPSON, J. C. 1979 An outbreak of suspected solanine poisoning in schoolboys: Examination of criteria of solanine poisoning. *Quarterly Journal of Medicine* **48**, 227–243.

MEEDENIYA, K. 1969 Investigations into the contamination of Ceylon desiccated coconut. *Journal of Hygiene, Cambridge* **67**, 719–729.

MELICK, C. O. 1917 The possibility of typhoid infection through vegetables. *Journal of Infectious Diseases* **21**, 28–38.

MELLING, J. & CAPEL, B. J. 1978 Characteristics of *Bacillus cereus* emetic toxin. *FEMS Microbiology Letters* **4**, 133–135.

MIDURA, T. F., SNOWDEN, S., WOOD, R. M. & ARNON, S. S. 1979 Isolation of *Clostridium botulinum* from honey. *Journal of Clinical Microbiology* **9**, 282–283.

MORSE, F. L. 1899 Report of medical inspector. *Massachusetts State Board of Health Report* **34**, 761–788. Cited by Geldreich, E. E. & Bordner, H. E. 1971 in *Journal of Milk and Food Technology* **34**, 184–195.

NOAH, N. D., BENDER, A. E., REAIDI, G. B. & GILBERT, R. J. 1980 Food poisoning from raw red kidney beans. *British Medical Journal* **281**, 236–237.

ODLAUG, T. E. & PFLUG, I. J. 1978 *Clostridium botulinum* and acid foods. *Journal of Food Protection* **41**, 566–573.

ORMAY, L. & NOVOTNY, T. 1969 The significance of *Bacillus cereus* food poisoning in Hungary. In *The Microbiology of Dried Foods* ed. Kampelmacher, E. H., Ingram, M. &

Mossell, D. A. A. pp.279–285. Proceedings of the Sixth International Symposium on Food Microbiology, Grafische Industrie Haarlem: The Netherlands.

OULSNAM, M. 1980 Microbiological condition of 1979 harvest flours. Flour Milling and Baking Research Association Bulletin No. 3, June, 95.

PATTERSON, J. A. 1972 Salmonellae in animal feedingstuffs. *Record of Agricultural Research* **20**, 27–33.

PIXLEY, C. 1913 Typhoid fever from uncooked vegetables. *New York Medical Journal* **98**, 328.

PORTNOY, B. L., GOEPFERT, J. M. & HARMON, S. M. 1976 An outbreak of *Bacillus cereus* food poisoning resulting from contaminated vegetable sprouts. *American Journal of Epidemiology* **103**, 589–594.

POWERS, E. M., LATT, T. G. & BROWN, T. 1976*a* Incidence and levels of *Bacillus cereus* in processed spices. *Journal of Milk and Food Technology* **39**, 668–670.

POWERS, E. M., LAWYER, R. & MASUOKA, Y. 1976*b* Microbiology of processed spices. *Journal of Milk and Food Technology* **38**, 683–687.

REBOLLO, M. R. 1964 Estudio sobre el contenido en enterobacterias de la "Lactuca sativa, L" (Lechuga). *Annales de Bromatologia* **16**, 395–404.

SAKAGUCHI, G. 1979 Botulism. In *Food-borne Infections and Intoxications* ed. Riemann, H. & Bryan, F. L. pp.390–442. London & New York: Academic Press.

SCHMITT, N., BOWMER, E. J. & WILLOUGHBY, B. A. 1976 Food poisoning outbreak attributed to *Bacillus cereus*. *Canadian Journal of Public Health* **67**, 418–422.

SEVERS, D. 1974 Salmonella food poisoning from contaminated white pepper. *Epidemiological Bulletin* **18**, 80.

SMART, J. L., ROBERTS, T. A., STRINGER, M. F. & SHAH, N. 1979 The incidence and serotype of *Clostridium perfringens* on beef, pork and lamb carcasses. *Journal of Applied Bacteriology* **46**, 377–383.

STEINIGER, F. 1957 Schnecken und Typhus — Paratyphosus — Bakterien (Snails and enteric bacteria). *Zeitschrift für angewandte Zoologie* **44**, 93–95. Abstracted in *Bulletin of Hygiene* 1957 **32**, 650.

SUGII, S. & SAKAGUCHI, G. 1977 Botulogenic properties of vegetables with special reference to the molecular size of the toxin in them. *Journal of Food Safety* **1**, 53–65.

TAMMINGA, S. K., BEUMER, R. R. & KAMPELMACHER, E. H. 1978 The hygienic quality of vegetables grown in or imported into the Netherlands: a tentative survey. *Journal of Hygiene, Cambridge* **80**, 143–154.

TERRANOVA, W., BREMAN, J. G., LOCEY, R. P. & SPECK, S. 1978 Botulism type B: Epidemiological aspects of an intensive outbreak. *American Journal of Epidemiology* **108**, 150–156.

VADHANASIN, S., POOMCHATRA, A., PAN-URAI, M. L. R., FOOPANICHPUCK, C. & THUDSRI, P. 1976 A bacteriological survey of foods from the flight and restaurant kitchens serving Bangkok International Airport. *Journal of the Medical Association of Thailand* **59**, 156–161.

VELAUDAPILLAI, T., NILES, G. R. & NAGARATNAM, W. 1969 Salmonellas, shigellas and enteropathogenic *Escherichia coli* in uncooked food. *Journal of Hygiene, Cambridge* **67**, 187–191.

VERNON, E. 1966 Food poisoning in England and Wales, 1965. *Monthly Bulletin of the Ministry of Health and the Public Health Laboratory Service* **25**, 194–207.

VERNON, E. 1967 Food poisoning in England and Wales, 1966. *Monthly Bulletin of the Ministry of Health and the Public Health Laboratory Service* **26**, 235–249.

VERNON, E. 1969 Food poisoning and *Salmonella* infections in England and Wales, 1967. *Public Health, London* **83**, 205–223.

VERNON, E. 1970 Food poisoning and *Salmonella* infections in England and Wales, 1968. *Public Health, London* **84**, 239–260.

VERNON, E. 1977 Food poisoning and *Salmonella* infections in England and Wales, 1973–75. *Public Health, London* **91**, 225–235.

VERNON, E. & TILLETT, H. E. 1974 Food poisoning and *Salmonella* infections in England and Wales, 1969–1972. *Public Health, London* **88**, 225–235.

WARRY, J. K. 1903 Enteric fever spread by watercress. *Lancet* **ii**, 1671.

WERNER DE GARCÍA, B., ARMANETTI DE STOTZ, E., SOUBES DE PESQUERA, N., CHAÍN, G. & DE MARCO DE CASTELÚ, N. 1978 Microbiología de vegetales: I Patógenos potenciales Gram negativos en lechuga escarola y berro. (Microbiology of vegetables. I Potentially pathogenic Gram negative micro-organisms in lettuce, endive and watercress). *Revista Latinoamericana de Microbiologia* **20**, 201-205.

WILSON, G. S. 1959 A small outbreak of solamine poisoning. *Monthly Bulletin of the Ministry of Health and the Public Health Laboratory Service* **18**, 207-210.

WILSON, M. M. & MACKENZIE, E. F. 1955 Typhoid fever and salmonellosis due to the consumption of infected desiccated coconut. *Journal of Applied Bacteriology* **18**, 510-521.

Bacteria in Frozen Vegetables

M. J. M. MICHELS

Unilever Research Laboratorium,
Vlaardingen, The Netherlands

Contents

1. Introduction

SINCE QUICK FROZEN vegetables first went on sale some 50 years ago, the consumer market for such products has expanded from a few basic vegetables like peas, beans and spinach to a range of over 25 different vegetables. At the moment, frozen vegetables form a major product group within most frozen food markets, with a *per capita* consumption ranging from 1–2 kg in Finland or Czechoslovakia to over 10 kg in the USA (Anon. 1977). Major products are maize and peas in the USA, green peas and green beans in the UK and France, and spinach in Germany, the Netherlands and Sweden. Recent developments on the market are ready-prepared vegetables in sauce and a range of vegetable mixes composed of common vegetables and more exotic ingredients like bean sprouts, red and green peppers and maize.

From the microbiological point of view, frozen vegetables can be divided into three different groups:

(1) *Blanched, frozen vegetables*

These are the common vegetables like spinach or peas, which are harvested, washed, blanched, cooled and frozen in retail packs.

(2) *Unblanched, frozen vegetables*

Some frozen vegetables are not blanched before freezing because this results in a better quality product. Examples are peppers, leeks and parsley.

(3) *Precooked, ready-to-serve vegetables*
These are vegetables which are part of quick-frozen meals: they are cooked before packing and generally hot-filled into the pack. Typical of this group are products with low bacterial counts, comparable to those of ready-prepared meals.

These groups are not always recognized in market products for three reasons. First, it is sometimes difficult to know whether a given frozen vegetable was blanched or not. Secondly, some vegetable mixes are composed of blanched and unblanched ingredients and therefore do not fit easily into this classification (although listed here in Group 2). Thirdly, there are comparable products, such as spinach in butter sauce and leek in butter sauce, which are produced by different processes. The spinach product is made from blanched and cooled spinach, which is mixed with a hot butter sauce, resulting in an end-product sometimes with an ideal temperature for bacterial multiplication, i.e. 20–40°C. The other product is prepared by directly mixing blanched leek, while still hot, with a hot sauce and the leek-in-sauce is then hot-filled. Thus, the spinach-in-sauce falls into Group 1; the leek-in-sauce, in Group 3.

One could argue that the subdivision proposed is of no value, when it is difficult to find out in which category a product falls, and therefore such a system does not help in the interpretation of the bacterial counts of frozen vegetables. To some extent this is true and the producer could possibly help in this respect by indicating whether the vegetables in the pack had been blanched or not. On the other hand, the bacterial counts of a product should always be judged in relation to the processing applied. So I would prefer to continue with this division into three major groups, the products of which have quite different bacterial loads.

The title of this contribution is 'Bacteria in Frozen Vegetables' and the main question could be: why are we interested in bacteria in frozen vegetables? The answer depends on who is asking the question. For the microbial ecologist, it is interesting to know which bacteria are present in the fresh vegetables, which bacteria are found on the processing line and which may be present in the frozen product. This has been a major topic since the start of this industry and many investigators have collected information on it. Another possible question may concern potential problems regarding the safety of the consumer. Early investigations were directed to *Clostridium botulinum* (James 1932), but now more interest is shown in salmonellas and other organisms (Anon. 1974*a*). The industrial food microbiologist wants to know whether a frozen vegetable is wholesome, safe for the consumer, and has been processed under Good Manufacturing Practices (GMP). Organisms of public health significance are discussed elsewhere in this symposium (Ch. 11, this volume): the observations in Section 3 relate therefore only to frozen vegetables. The relationship between GMP and the bacterial counts of vegetables will be discussed separately for each of the three groups of frozen vegetables.

Finally, a survey of bacterial counts of some frozen vegetables will be given, and problems related to industrial or legal maximum bacterial counts will be discussed.

2. Composition of the Bacterial Flora

In an early paper on the microbiology of frozen food it was stated that the bacteria on raw vegetables originate from the soil and from vegetable litter (James 1932) and, as expected, bacteria in frozen vegetables are mainly of the common soil types (in the genera *Achromobacter*, *Bacillus*, *Flavobacterium* and *Pseudomonas*). Additionally, there are other sources of bacterial contamination such as washwater, containers and the operators themselves (Smart & Brunstetter 1937). An early observation was that blanching caused quite a change in composition of the bacterial flora, from *Achromobacter*, *Pseudomonas* and *Flavobacterium* characteristic of the fresh vegetables to *Bacillus*, *Flavobacterium* and *Sarcina* (Smart 1937). Frozen storage also caused a change in composition of the microflora with better survival of fairly freeze-resistant micrococci and flavobacteria, and a rapid decrease of *Achromobacter* species (Lochhead & Jones 1936).

The microbial flora is also affected by the vegetable itself, e.g. peas, on which mainly *Leuconostoc* and *Streptococcus* species were found (White & White 1962; Cavett *et al.* 1965), and also by the length of time a processing line is in operation (Splittstoesser & Gadjo 1966). It is obvious, therefore, that it is difficult to speak of *the* bacterial flora of frozen vegetables: nevertheless, Table 1 lists bacterial groups that were frequently isolated from frozen vegetables.

An interesting observation regarding the above-mentioned effect of blanching on the composition of the bacterial flora was made by Splittstoesser & Gadjo (1966). They found mainly spore-forming bacteria directly after blanching, but because of re-contamination during further processing, the bacterial flora of post-blanch samples of peas and beans was similar to that of the raw vegetables. No studies have been made on the composition of the bacterial flora of unblanched frozen vegetables, but one may expect this flora to resemble that of the raw vegetable, because these foodstuffs will not have been subjected to processing steps likely to have a significant effect on the bacteria present.

TABLE 1

Bacterial genera frequently isolated from frozen vegetables

Genus*	Frozen vegetable	Reference
Bacillus, *Achromobacter*, *Micrococcus*	Spinach, peas	Brown (1933)
Sarcina, *Flavobacterium*, *Achromobacter*, *Bacillus*, *Lactobacillus*, *Pseudomonas*, *Streptococcus*, *Aerobacter*	Green beans, beets, corn, Lima beans, mushrooms, peas, tomatoes, spinach	Smart (1934)
Micrococcus, *Flavobacterium*, *Achromobacter*, *Bacillus*	Asparagus, peas, beans, corn	Lochhead & Jones (1936)
Bacillus, *Flavobacterium*, *Micrococcus*, *Pseudomonas*	Kale, spinach	Smart & Brunstetter (1937)
Bacillus, *Flavobacterium*, *Lactobacillus*, *Sarcina*, *Micrococcus*	Peas, green beans, Lima beans, corn	Smart (1937)
Micrococcus, *Flavobacterium*, *Bacillus*, *Achromobacter*	Asparagus, spinach, beans, peas, corn	Lochhead & Jones (1938)
Sarcina, *Flavobacterium*, *Bacillus*, *Achromobacter*, *Pseudomonas*, *Lactobacillus*	Green beans, wax beans, Lima beans, mushrooms, spinach, peas	Smart (1939)
Micrococcus, *Flavobacterium*, *Achromobacter*, *Streptococcus*	Peas, beans, corn	Hucker *et al.* (1952)
Leuconostoc, *Streptococcus*	Peas	White & White (1962)
Leuconostoc, *Streptococcus*	Peas	Cavett *et al.* (1965)
Streptococcus, *Leuconostoc*, *Micrococcus*, *Achromobacter*, *Pseudomonas*, *Flavobacterium*, *Corynebacterium*, *Arthrobacter*, *Microbacterium*	Green peas, snap beans, whole-kernel corn	Splittstoesser & Gadjo (1966); Splittstoesser (1970, 1973)

*Where possible, most frequently isolated genera are listed first.

3. Organisms of Public Health Significance

Probably the first reports on *Salmonella* in frozen vegetables are those of Lochhead & Jones (1936, 1938), who found 4·7% of colonies isolated from 'fresh pack' corn and 2·5% of isolates from 'fresh pack' asparagus to be *Salmonella* (from their papers it is not clear whether these 'fresh pack' samples were taken just before or directly after freezing). More recent reports describe the finding of salmonellas in green beans (Raccach *et al.* 1972) and in corn kernels (Sadovski 1977), both in samples taken before blanching from freezing plants in Israel. Other reports point to the presence of salmonellas in a number of vegetables from the market, such as lettuce and fennel in Italy (Ercolani 1976) and in a range of—mainly imported—vegetables in the Netherlands (Tamminga *et al.* 1978). *Salmonella* contamination of fresh vegetables has led to a number of food poisoning outbreaks, as cited by Geldreich & Bordner (1971), but such outbreaks seem not to be caused by salmonellas from *frozen* vegetables. In effect, there are no recent reports on the presence of salmonellas in such vegetables, although one should realize that the number of samples investigated is limited: 61 samples of Brussels sprouts, carrots, peas, spinach, curly kale and broad beans (Wagner & Borneff 1967*a*) and 47 samples of peas, green beans, Brussels sprouts, carrots, spinach, asparagus, corn and lima beans (Splittstoesser & Segen 1970). It was concluded by Splittstoesser & Segen (1970) that vegetable processing lines do not provide a favourable environment for salmonellas and that they would be inhibited by the normal microflora of frozen vegetables. This was not the finding of White & White (1962), who demonstrated that salmonellas (and staphylococci) developed well in blanched peas. They concluded that the normal flora would not prevent the growth of pathogens in food. Probably the absence of salmonellas in frozen blanched vegetables can best be explained on the basis of a low *Salmonella* incidence, reduction by washing, elimination by blanching and—in the case of recontamination—reduction by freezing and frozen storage. The effect of freezing and frozen storage has been studied in model experiments by Borneff & Wagner (1967). They observed a decrease by a factor of 10 of salmonellas in spinach by freezing alone, and further reductions after frozen storage.

For unblanched, frozen vegetables, one may expect a very low incidence of salmonellas, because of some of the factors already mentioned, namely, a low incidence, further reduced by washing, freezing and frozen storage. Data available are, however, very limited. Käferstein (1976) found no salmonellas in 14 samples of frozen parsley, and we could not detect them in 36 samples of (mainly) leeks, peppers and onions.

There are some reports on staphylococci in frozen vegetables. In an investigation by Jones & Lochhead (1939) 46% of all isolates from frozen asparagus, spinach, peas, beans and corn were micrococci and about half of these micrococci were enterotoxic staphylococci. This remarkably high incidence was not observed by Splittstoesser *et al.* (1965) who detected staphylococci only at a low level (1·4–7·3/g) in 112 samples of peas, green beans and corn. Similar results were obtained by Wagner & Borneff (1967*a*), with <10/g in 61 samples, by Reeves (1973) with only eight of 234 samples with 10 staphylococci/g, and in a 1979 Dutch survey (B. J. Hartog, pers. comm.) with 99% of 350 samples having <100/g. From the latter data one may therefore conclude that staphylococci are of little significance in frozen vegetables.

4. Effects of Processing on Bacterial Counts of Blanched Vegetables

Vegetables are generally harvested, transported to the processing factory, cleaned, roots or other inedible parts removed, thoroughly washed, blanched and cut or chopped. Then they are either filled in consumer packs and frozen, or frozen before packing, stored in bulk and packed later. Of all these processing steps, blanching has the greatest effect on the bacterial load of the vegetables. The blanching is carried out by passing a vegetable by conveyor belt or screw through a bath of water at a temperature of 95–99°C, with a residence time of 1–5 min. The main objective of this heat treatment is to inactivate tissue enzymes which could adversely affect the quality of the product, even when stored at –20°C.

Blanching potentially eliminates all vegetative bacteria from the fresh vegetable, leaving only bacterial spores in the blanched product. In practical terms, this means that the bacterial load of the fresh vegetable can be reduced from $1 \times 10^6 - 20 \times 10^6$ to 10^2–10^4/g, a reduction of 99·9% (Smart & Brunstetter 1937; Jones & Ferguson 1956). Although reduction levels of *ca.* 80–99% are also quoted (Smart 1937; Pederson 1947; Wagner & Borneff 1967*b*) and counts may well exceed 10^4/g after blanching (Mohs 1969), it is our experience that aerobic plate counts directly after blanching range from 10^2 to *ca.* 10^3/g, which are reductions of 99·9% or more.

A processing step of which the effect is sometimes overestimated is the washing of a vegetable. Thorough washing of the vegetables, sometimes by two or three washing stages, removes soil and dirt, but the effect on the bacterial load, however, is limited, reducing the bacterial numbers by only 30–80% (Jones & Ferguson 1956; Wagner & Borneff 1967*b*; Mohs 1969). The processing steps after blanching usually result in some increase in the microbial load of the vegetables, and it is the responsibility of the processor to conduct operations in such a way that they cause no avoidable increase in the number of bacteria. Nevertheless, currently used equipment and processing often involve the use of water for cooling or transport, open belts for transportation and an environment in which re-contamination is unavoidable. Ways of reducing post-blanching contamination in the factory were indicated 35 years ago by Vaughn *et al.* (1946), who recommended: straight in-line processing, better design of hygienic equipment, continuous cleaning of equipment, use of potable water and in-plant freezing. Most of these recommendations are followed nowadays, although there is still a need for improved design of equipment allowing simpler and easier cleaning and less build-up of contamination. A significant improvement has been made in recent years in the freezing of vegetables, which formerly was done batchwise and allowed some multiplication of bacteria before freezing actually took place. Nowadays, freezing is generally an in-line operation, where the use of automatic plate freezers or fluidized-bed freezers has eliminated holding times before freezing.

In connection with the levels of bacteria which may be found on frozen vegetables, it is profitable to discuss one example from the literature regarding processing of French-style beans and, as another example, the processing of spinach by two different routes.

In an investigation by Splittstoesser *et al.* (1961*a*) French-style beans had a low aerobic plate count of 380/g directly after blanching (see Fig. 1).

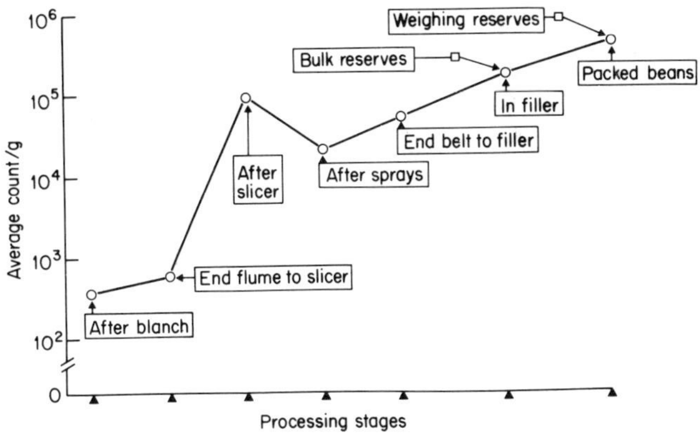

Fig. 1. Microbial contamination of French-style beans during processing (based on Splittstoesser *et al.* 1961*a*).

Passing the beans through a flume to the slicer resulted in a slight increase, but the slicing itself caused a sudden increase in bacterial counts to a mean value of almost 10^5/g. Watersprays reduced this contamination, but in the filler, a further increase was observed because some batches of beans were stored as bulk reserves and were added later on to the beans from the belt to the filler. A further increase in bacterial numbers was caused by weighing reserves, i.e. beans used to correct any underweight in machine-filled boxes. The mean count for the material thus added was close to 10^6/g and it was assumed that bacterial multiplication in the beans caused this high level of contamination. This example highlights some of the causes of contamination of blanched vegetables. Improved design and/or cleaning of the slicing machine should be encouraged to eliminate this additional source of contamination. The contamination due to bacterial growth in the weighing reserve could be eliminated by freezing the sliced beans in a fluidized-bed freezer and using frozen beans as weighing-in material. For beans and peas such freezing before packing was shown to reduce the bacterial count of the end product (Mohs 1969).

Traditionally, spinach is machine-harvested and transported to the freezing plant, where it is washed thoroughly, blanched and cooled directly in water. Thereafter, the water is removed, the spinach is minced and pumped via a buffer tank to filling machines, where it is packed in consumer packs of different sizes. Formerly, the packs were batch-frozen in a plate freezer but today this is increasingly done as an in-line operation via automatic plate freezers. This process with direct water-cooling of the spinach requires much water, and therefore a different procedure was developed some years ago in which the spinach was not cooled after blanching, but was minced while still hot, followed by indirect cooling of the minced spinach via a double-tube heat exchanger. Then the spinach was filled in the usual way into consumer packs and frozen. Table 2 gives the counts of bacteria in spinach, processed by direct cooling of the blanched material. In the Netherlands the use of Enterobacteriaceae counts is common, whereas in other countries coliforms are often estimated, so both counts are included in the Table. As mentioned earlier, fresh spinach can harbour high bacterial levels which are reduced to below 10^3/g by blanching.

The processing steps after blanching are likely to increase the bacterial load. We have taken samples at the start of the processing season, when the line was completely clean, and from the line in the high season, when bacterial build-up had occurred and cleaning done with minimal time loss. Although the columns of Table 2 show that with the direct cooling system under optimal conditions, low counts of Enterobacteriaceae and coliforms could be achieved, it is nevertheless clear that higher counts may occur because of re-contamination via partly re-cycled cooling water, from bacteria in boiler scale in water separators, from inadequate cleaning of equipment or transport belts, and from the open layout of the line. Moreover, the cooled spinach at about 20°C was chopped, transported and filled at ambient temperatures, which allowed bacterial multiplication.

TABLE 2
Examples of bacterial counts of spinach (with direct cooling system)*

Treatment	Standard Plate Count		Enterobacteriaceae		Coliforms	
	(a)	(b)	(a)	(b)	(a)	(b)
Fresh, before washing	—	$8{\cdot}6 \times 10^6$	—	$1{\cdot}2 \times 10^6$	—	$7{\cdot}3 \times 10^5$
After washing	—	$8{\cdot}2 \times 10^5$	—	$1{\cdot}1 \times 10^5$	—	$5{\cdot}1 \times 10^4$
After blanching	$5{\cdot}7 \times 10^2$	$4{\cdot}7 \times 10^2$	<10	<10	<10	<10
After water separator	$6{\cdot}6 \times 10^3$	$2{\cdot}4 \times 10^3$	<10	$1{\cdot}8 \times 10^3$	<10	$9{\cdot}7 \times 10^2$
After chopper	$1{\cdot}3 \times 10^4$	$5{\cdot}9 \times 10^4$	80	$2{\cdot}0 \times 10^3$	70	$1{\cdot}4 \times 10^3$
Filled in pack	$1{\cdot}4 \times 10^4$	$2{\cdot}1 \times 10^5$	70	$6{\cdot}3 \times 10^3$	30	$5{\cdot}2 \times 10^3$

*All counts expressed as number/g of product.
(a) Start of season, no bacterial build-up.
(b) High season, with bacterial build-up.

TABLE 3
Examples of bacterial counts of spinach (with indirect cooling system)*

Treatment	Temperature (°C)	Standard Plate Count	Enterobacteriaceae	Coliforms
Fresh, before washing	20	$4{\cdot}0 \times 10^6$	$7{\cdot}8 \times 10^4$	2×10^4
After washing	20	$2{\cdot}9 \times 10^5$	1×10^3	3×10^2
After blanching	90	$1{\cdot}2 \times 10^3$	<10	<10
After chopper	70	$2{\cdot}6 \times 10^3$	<10	<10
After cooling	20	$2{\cdot}2 \times 10^3$	10	<10
Filled in pack	20	$2{\cdot}7 \times 10^4$	10	<10
Frozen product	-20	$2{\cdot}9 \times 10^4$	10	10

*All counts expressed as number/g of product.

Table 3 gives temperature and bacterial counts of spinach chopped while hot and cooled via the indirect cooling system. Because of the high temperature of the chopped spinach, which eliminated multiplication of Enterobacteriaceae organisms or coliforms, and indirect cooling, which prevented re-contamination via the cooling water, the spinach produced had low bacterial counts. In practice, the low bacterial levels shown in Table 3 are not always realized and higher counts were found by Wagner & Borneff (1967*b*), who observed a standard plate count of $>10^5$/g (mainly diplococci) in spinach processed via the indirect cooling system. They suggested that a 'total' viable count determination would be sufficient for the routine quality control of the process. We, on the other hand, prefer a count of Enterobacteriaceae or coliforms as this test allows corrective action to be taken within a day, whereas a standard plate count requires two or three days. Secondly, a test for Enterobacteriaceae or coliforms can yield more useful information than a standard plate count, as one can see from a comparison of columns (a) and (b) in Table 2.

The main objective of this section is to give some insight into the factors that may affect the bacterial counts of blanched vegetables, as these counts are largely a reflection of the processing system employed. Therefore, proper interpretation of bacterial count data requires detailed information on the processing involved for the vegetable under study. Data are given here as examples and further information can be found in the literature, where Pederson (1947), Hucker *et al.* (1952), Jones & Ferguson (1956), Splittstoesser *et al.* (1961 *a,b*), Splittstoesser & Gadjo (1966), Mohs (1969) and Raccach *et al.* (1972) discuss the processing of peas, beans, corn and spinach.

5. Effects of Processing on Bacterial Counts of Unblanched Vegetables

In the last few years, several kinds of unblanched, frozen vegetables and herbs have appeared on the market. Blanching does not improve the quality of frozen: cabbage, carrot, celery root, corn, leek, onion, green and red pepper, tomato, basil, celery leaf, chervil, chives, cress, dill, parsley or tarragon; indeed it causes loss of flavour and/or deterioration of texture. For this group of vegetables, the omission of blanching gives a product which, even after storage for *ca.* 1 year, will still be better than the blanched counterpart (Kozlowski 1977). Some of these vegetables and herbs, such as parsley, dill, chive and onions, are sold as such to the consumer. Others like mushrooms, leeks and peppers are used as ingredients in other frozen foods such as soup greens, vegetable mixes or pizzas.

Except for the blanching step, these vegetables are processed like blanched frozen vegetables, i.e. they are harvested, transported, peeled and/or cut, washed and frozen. The freezing is commonly done in fluidized-bed freezers resulting in a 'free-flowing', individually quick-frozen product, which can easily be used by the consumer or mixed industrially with other vegetables.

From the microbiological point of view, these vegetables pose some interesting new questions, as it is no longer possible to judge the degree of Good Manufacturing Practice applied during processing by determining Standard Plate Counts or numbers of Enterobacteriaceae/coliforms. This is because most fresh vegetables contain 10^4–10^6 Enterobacteriaceae/g and have Standard Plate Counts of 10^5–10^7/g. Washing and cleaning the vegetables removes part of the bacterial flora, but the high and variable numbers of bacteria that remain account for the lack of correlation between Standard Plate Counts or counts of Enterobacteriaceae/coliforms and the GMP applied in processing.

Unblanched frozen vegetables are therefore distinctly different microbiologically from blanched vegetables, which leads to the following two questions. How do we know whether a vegetable has been blanched or not? Secondly, which analytical techniques are available to monitor or judge the condition of an unblanched vegetable? For a microbiologist in a processing plant the first question is easy to answer because he knows the situation, but a microbiologist from an outside controlling agency might have a problem in finding out. There are, however, two simple enzyme tests, in which one tests for the presence of catalase with hydrogen peroxide, or for peroxidase with guaiacol and hydrogen peroxide. Details of these tests are described elsewhere (Luh & O'Neal 1975). For our purposes a simplified peroxidase test may be applied, in which three drops of 1% guaiacol in water and three drops of 3% H_2O_2 are added to a piece of the vegetable under test. No colour change within 1 min is a negative reaction (blanched vegetable), whereas a pink-brown discolouration indicates an unblanched sample (Anon. 1979).

In cases of doubt, application of these tests will help to decide whether a vegetable has been blanched or not, and will prevent the misinterpretation of laboratory findings, such as high counts of Enterobacteriaceae. The second problem concerning suitable microbiological tests for unblanched vegetables is more difficult to solve. One could choose to test for salmonellas, but as indicated earlier, the incidence of salmonellas in raw vegetables is so low, that a routine analysis is not warranted. Probably a more suitable microbiological test for unblanched vegetables is to estimate *Escherichia coli* as a test for potential faecal contamination of the product. It is important to stress the word 'potential' because *E. coli*, in our experience, grows well in vegetable material, so an elevated *E. coli* count in the frozen product could be caused by bacterial multiplication somewhere on the processing line, and is not necessarily a direct indication of severe faecal contamination with a concurrent risk of an increased incidence of *Salmonella*, as was found by Tamminga *et al.* (1978) for unprocessed vegetables from the market.

Finding no *E. coli* (or *Salmonella*) in an unblanched vegetable does not necessarily mean that the freezing line, over which the product passed, was sufficiently clean and run hygienically. We have wondered whether another group of organisms might be indicative of the effectiveness of the GMP applied. We have no firm answer to this question yet, but we are paying more attention to lactobacilli because some years ago we noticed that these organisms were almost absent (<10/g) from freshly washed vegetables from the market, whereas levels of 10^3–10^4/g were found in their ready-to-cook counterparts, which were sold from the chilled cabinet with a shelf-life of a few days. We subjected these vegetables (onions, carrots, leeks and endive) to a number of tests: Standard Plate Counts and counts of Enterobacteriaceae, *Pseudomonas*, lactobacilli, yeasts and moulds, and observed distinct differences only for the lactobacilli (Fig. 2) in the products from the chilled cabinet compared with their freshly prepared counterparts.

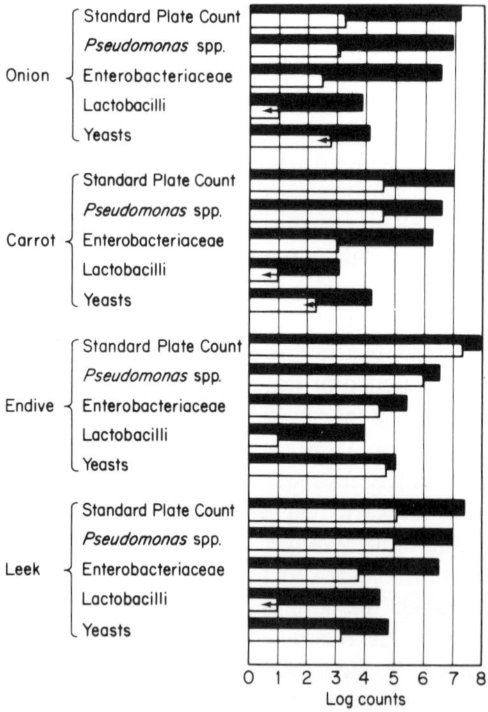

Fig. 2. Means of bacterial counts for washed and sliced vegetables. □ Fresh, domestically prepared (mean of 3 samples); ◀—┤, less than; ■ commercial, ready-to-cook (mean of 7–12 samples).

As an example the bacterial counts of unblanched leeks, including a count of lactobacilli, are given in Table 4. Counts of Enterobacteriaceae and the Standard Plate Count are included to emphasize the point already made about the limited significance of these tests. Processing of the leeks involved harvest, transport to the freezing plant in boxes, removal of the roots, coarse cutting, washing, a second final cut, a second washing and belt transport to the freezer where the pieces of leek were individually quick-frozen within minutes in a fluidized-bed freezer. The frozen product was stored in bulk containers for later use in vegetable mixes.

The bacterial counts in Table 4 show that the leeks entered the factory with a high natural load of bacteria including Enterobacteriaceae, and

TABLE 4
*Bacterial counts of unblanched leek**

Processing stage	Standard plate count ($\times 10^6$/g)	Enterobacteriaceae ($\times 10^5$/g)	Lactobacilli (per g)
Fresh, derooted	3–10	3–10	<100–900
Cut and washed†	4	2	<100
After second cut	4–50	4–50	100–1200
Twice cut and washed†	1–10	1–6	<100–1000
Before froster	1–7	0·6–2	<100–4000
Frozen after froster	2–10	0·6–2	<100–4000

*Range of counts from three investigations on a processing line. *Salmonella* spp. (in 25 g) and *E. coli* (in 0·1 g) were not detected in 12 samples of roots and six samples of frozen leek.
†Bacterial counts of wash water during processing were *ca.* 10 times lower than that of leek passing through, indicating an adequate flow rate of fresh water.

processing had little effect on these counts. It is noteworthy that the freezing had no direct adverse effect on the counts of the Enterobacteriaceae, which normally are fairly sensitive to freezing. The explanation might be that the rapid freezing involved was close to the optimal freezing speed for preservation of cells, i.e. close to 7°C/min. Referring further to the lactobacilli, their level was low in the leeks entering the plant ($<10^2$ – *ca.* 10^3/g) and processing itself gave only an insignificant increase (to $<10^2$ – 4×10^3/g). Routine observation of the level of lactobacilli in a few hundred samples of the frozen leeks resulted in 29% with $<10^3$/g, 60% with 10^3 – 2×10^4/g and 11% with 2×10^4 – 10^5/g. We think that levels well over 10^5/g, which occur occasionally, might indicate substandard raw material or insufficient hygiene on the line.

As obvious dirty areas were not present in the processing line we examined, we analysed a few samples of product resting at positions along the line to find out whether lactobacilli could be considered part of the spoilage flora. In some samples (like vegetable sludge) lactobacilli were present at a level of 10^5 – 10^6/g. Comparable levels were found in frozen leek incubated for 20 h at ambient temperatures. From these results we deduced that lactobacilli could be involved in spoilage of unblanched leek or other vegetables. This is in agreement with the observations by Berry (1933) on spoilage of defrosted peas, beans and corn. Mundt (1970) also referred to millions of lactobacilli in expressed vegetable juices on processing equipment. These findings, together with the naturally low incidence of lactobacilli on plants, from less than 10 to a few hundred per gram (Mundt & Hammer 1968; Mundt 1970)—could make testing for lactobacilli relevant for unblanched vegetables.

At present, insufficient data are available to draw definite conclusions about elevated lactobacilli levels; furthermore the incubation time required (3 d at 37°C or 5 d at 30°C, using Rogosa agar) makes this test somewhat unsuitable for the rapid detection of incidental deviations from otherwise good manufacturing practice. However, in cases of consistently high levels of lactobacilli, it would be worthwhile to find out whether the lactobacilli originate from the raw material or from the processing, for these lactobacilli might be a clue to substandard conditions.

6. Effects of Processing on Bacterial Counts of Prepared Vegetables

The cooked vegetables filled hot into trays (our processing type Group 3) form an integral part of frozen dinners, discussion of which falls outside the scope of this paper. These vegetables are mentioned because others, the prepared vegetables, which are not always fully cooked, may have bacterial counts quite close to those of cooked vegetables. Examples are butter beans in Sauce Provençal and red cabbage with apples. The latter can be made from hot, sliced and cooked cabbage to which hot apple slices are added. The mixture is then filled hot into the package and frozen. This acid vegetable product may have Standard Plate Counts below 10^3/g in 95% of samples, and Enterobacteriaceae <10/g in 99% (based on 300 samples): such levels are even lower than those commonly found for frozen dinners, probably because of the pH of 5·0.

The butter beans in sauce may be processed differently by cooking the beans, cooling them in water, followed by fluidized-bed freezing. The resulting free-flowing frozen beans can then be packed frozen, with the concurrent addition of the hot sauce. Subsequently the resulting vegetable product is frozen. Because water-cooling of the cooked beans causes some re-contamination, bacterial counts of this product are, from our own experience, slightly higher than for the other product, i.e. Enterobacteriaceae were <10/g in 88% of samples, 10–10^2/g in 9% and 10^2–10^3/g in 4% (based on 57 samples). Standard Plate Counts of 10^3–10^5/g may be expected, although lower as well as higher levels do occur. As mentioned in the introduction, not all prepared vegetables have counts close to those of cooked vegetables; depending on the processing involved, bacterial counts may be the same or slightly higher than for the blanched vegetables. For example, spinach in butter sauce is made by adding 10% of hot butter sauce to spinach commonly blanched and cooled. This increases the temperature of the spinach to about 30°C (ideal for bacterial multiplication) and together with the additional mixing and pumping required, may cause the bacterial levels to rise above those found for straight chopped spinach: the obvious alternative of mixing hot sauce with hot, blanched spinach is not feasible, because of the sensitivity of hot spinach to discolouration.

The conclusion from this section is that bacterial counts of ready-prepared vegetable products may range from those found for the cooked, frozen dinners to those for the commonly blanched vegetables. The processing involved cannot always easily be deduced from the form in which the product is presented (compare the processing of leek and spinach in butter sauce referred to in the introduction), and in case of doubt, one should try to elucidate the processing involved, because such information may be vital for the correct interpretation of Standard Plate Count and Enterobacteriaceae/coliform count data.

7. Bacterial Counts of Blanched Vegetables

There are many reports on the microbiology of frozen vegetables, but surveys on the bacterial counts of blanched vegetables are limited. Examples are those of Jones & Ferguson (1956) for 220 samples, of Wagner & Borneff (1967*a*) for 61, and of Reeves (1973) for 234 samples. The latter found mean Standard Plate Counts from $3{\cdot}4\times10^3$/g for carrots to $4{\cdot}7\times10^6$/g for broccoli, which is in agreement with earlier observations on the variable counts of blanched vegetables. Standard Plate Counts of over 10^6/g were observed in *ca.* 20% of the samples by Jones & Ferguson (1956) and Wagner & Borneff (1967*a*).

TABLE 5

Percentage distribution of counts of bacteria in chopped spinach and spinach in butter sauce (creamed spinach)

Standard Plate Count			Count of Enterobacteriaceae*		
Count/g	Distribution (%) of counts in		Count/g	Distribution (%) of counts in	
	Spinach (1500)†	Creamed spinach (320)		Spinach (2700)	Creamed spinach (600)
$<10^4$	50	32	<100	79	57
10^4–10^5	33	47	10^2–10^3	17	30
$10^5 - 5\times10^5$	15	14	10^3–10^4	3	10
$5\times10^5 - 2\times10^6$	2	6	$10^4 - 5\times10^4$	1	2
$>2\times10^6$	0	1	$>5\times10^4$	0	1

*Enumeration without previous resuscitation.
†Number of samples analysed given within brackets.

Table 5 presents counts of chopped spinach and of chopped spinach in butter sauce, the former being the main frozen vegetable in the Netherlands. The Table shows that the addition of sauce required for the latter product cause a slight shift to higher counts. Data are from one processor applying direct and indirect cooling of spinach. Further data are given for a large processor and for combined large plus small processors (Table 6). The latter set of data shows a slight shift to higher counts, probably because of the less

TABLE 6

Percentage distribution of counts of bacteria in frozen vegetables: results from a large processor alone and combined results from large plus small processors

Standard Plate Count			Count of Enterobacteriaceae*		
Count/g	Distribution (%) of counts from		Count/g	Distribution (%) of counts from	
	Large processor (4600)	Large + small processors (2200)		Large processor (8200)	Large + small processors (3700)
$<10^4$	53	10	$<10^2$	78	59
10^4–10^5	27	42	10^2–10^3	17	30
$10^5 - 5 \times 10^5$	17	38	10^3–10^4	4	9
$5 \times 10^5 - 2 \times 10^6$	3	10	$10^4 - 5 \times 10^4$	1	1
$>2 \times 10^6$	0	0	$>5 \times 10^4$	0	1

*Enumeration without previous resuscitation.
Number of samples analysed given within brackets.

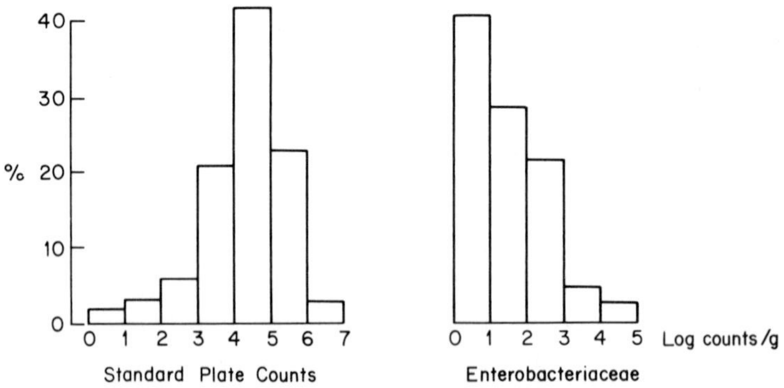

Fig. 3. Percentage distribution of counts for blanched frozen vegetables (298 samples) (B. J. Hartog, pers. comm.).

well-developed technical facilities and microbiological expertise of the smaller processors. Comparable further data on blanched, frozen vegetables are shown in Fig. 3, which is based on unpublished data from a 1979 survey of products on the Dutch market by Food Inspection Services and the Public Health Laboratory (B. J. Hartog, pers. comm.).

8. Bacterial Counts of Unblanched Vegetables

As discussed earlier, Standard Plate Counts and counts of Enterobacteriaceae are of little significance for unblanched vegetables. Nevertheless, to give an indication of the levels one can expect, Standard Plate Counts of leek are presented. Of 240 samples, 22% had counts of $<10^5$/g, 57% had 10^5–10^6/g and 21%, 10^6–10^7/g. Tests for *E. coli* with the direct plate method of Anderson & Baird-Parker (1975) showed *E. coli* to be absent (<20/g) in 99% of 1000 samples of unblanched vegetables and of mixes containing such vegetables. This low detection level was not caused by intrinsic inhibitive properties of onions, leek or peppers, because no inhibition of growth of *E. coli* was observed when it was inoculated on tryptone–bile agar–membrane plates, to each of which 1 ml of a 1/10 dilution of these vegetables had been added.

For the prepared vegetables, no survey data are given, because, as already discussed, counts may vary greatly, and some examples of typical counts have already been given.

9. Significance of Bacterial Counts of Frozen Vegetables

An early observation by Smart (1934) on strawberries was that a low microbial count did not necessarily mean a high-grade frozen product. This was confirmed by Pederson (1947), who found that of four lots of peas, the good quality peas had high counts (*ca.* 2×10^5/g), whereas the poor quality peas had much lower counts (*ca.* 9×10^3/g). The explanation given was that the superior peas were tender and sensitive to damage and therefore favoured subsequent bacterial growth, whereas the inferior peas which were over-mature and not suitable for freezing did not support such growth. Studies by Hucker *et al.* (1952) also led to the conclusion that there was no pronounced relationship between numbers of bacteria present and the quality of frozen vegetables. The lack of correlation between numbers of bacteria present and quality of the product is quite understandable from the data of Pederson (1947) who found 10^7–10^9 bacteria/g in spoiled peas. Frozen vegetables have counts well below these levels and so counts are therefore insignificant as indicators of flavour or quality.

Tressler (1938) pointed out the value of the bacterial count as an excellent indication of the kind of handling that a frozen vegetable had undergone. He stated that a count below 8×10^4/g would indicate proper blanching, proper cooling, clean equipment and rapid freezing, but he failed to suggest an interpretation of the commonly occurring higher counts. A similar, but more detailed, view was taken by Pederson (1947), who concluded that bacterial counts should be seen as an indication not of quality, but of the handling of the vegetable between blanching and final freezing. Very low counts indicated proper blanching and extremely clean equipment: very high counts, $>10^6$/g, were considered to result from contamination of equipment and an overlong holding period before freezing.

In present-day terms, Pederson realized that bacterial counts indicate the grade of Good Manufacturing Practice applied. In the past, a range of maximal bacterial levels for blanched vegetables produced under GMP conditions, was proposed. Most of these limits were defined when knowledge about the distribution of bacteria in food was scarce and the values proposed for counts were not always realistic. Much has improved since then due to the activities of the International Commission on Microbiological Specifications for Foods. But in the meantime, it has been realized that it is irrelevant to lay down legal standards for foods which already have an excellent public health record. Nevertheless, to indicate current industrially recognized bacterial limits, two specifications are given (Table 7) which are in use in Israel (Anon. 1974*b*) and France. The French values form part of governmental buying specifications for vegetables for public institutions. These limits have not yet been accepted fully by the French industry which considers the specifications to be very strict (Anon. 1979). This is probably not because of the limits for Standard Plate Counts and coliform counts, which are fairly realistic, if one takes into account the enumeration tolerance levels allowed and the higher limits for additionally processed vegetables such as chopped spinach. The problem seems to be the number of additional criteria which, with the possible exception of the *Salmonella* limit, are too strict, or irrelevant, or both. To a lesser extent this is true also for the Israeli standard, which includes a limit for enterococci, which are considered by many not to have any hygienic significance for frozen vegetables (Splittstoesser & Gadjo 1966; Mundt 1976). For blanched vegetables the above-mentioned *Salmonella* limit—and certainly the one for *Staphylococcus aureus*—is also questionable, because testing for these organisms is likely to result in almost exclusively negative findings. This view is held also by the ICMSF (Anon. 1974*a*), which suggests standards only for blanched frozen vegetables, on the basis of Standard Plate Counts and coliform counts. The limits proposed by the ICMSF—which are now under review—are, however, too strict, which emphasizes once more that the defining of microbiological specifications is a difficult matter.

TABLE 7

Israeli and French specifications for frozen vegetables, expressed as maximum values for counts/g

Type of count	Specification for Israel (in 1974)	France (in 1979) A	France (in 1979) B
Standard Plate Count	$<5 \times 10^5$	$<5 \times 10^5$ *	$<1{\cdot}5 \times 10^6$ *
Coliforms	$<5 \times 10^2$	$<10^3$ *	$<3 \times 10^3$ *
Faecal coliforms	—	<15*	<15*
Enterococci	$<10^3$	—	—
Spores of sulphite-reducing clostridia at 46°C	—	<10*	<10*
Spores of *Clostridium perfringens*	—	<1*	<1*
Staphylococcus aureus	$<10^2$	<10	$<10^2$
Salmonella spp.	Absent (in 50 g)	Absent (in 25 g)	Absent (in 25 g)
Yeasts	—	$<10^3$ *	$<2 \times 10^3$ *
Moulds	—	$<5 \times 10^2$ *	$<1{\cdot}5 \times 10^3$ *

A, frozen vegetables; B, frozen vegetables with additional processing between blanching and freezing, e.g. chopped spinach.
*To allow for analytical errors maximum values listed may be exceeded by a factor of 3 (using solid media) or 10 (using liquid media).

Two other factors complicate the situation even further. Bacterial counts of frozen vegetables are not stable, but are reduced during frozen storage. This was known by Lochhead & Jones (1936) who observed a reduction by a factor of 7–20 for Standard Plate Counts of vegetables stored for six weeks at −18°C. Such reductions complicate the recommendation of specifications when data are derived from surveys of products taken from the market; the specifications should be applicable to the freshly produced product as well. The other complicating factor is that some of the bacteria present are damaged by freezing and subsequent frozen storage. Proper enumeration of Enterobacteriaceae or coliforms requires resuscitation of damaged cells in a non-selective medium before any selective medium is used. One way of doing this is by pre-incubating a 1/10 dilution of the sample in Trypticase Soy Broth for 1 h at 20°C. An even simpler procedure, i.e. resuscitation in the commonly used peptone—salt diluent, often gives significantly increased Enterobacteriaceae counts. This was shown for frozen vegetables, which were kept for 16 h at ambient temperature prior to freezing to increase the Enterobacteriaceae level (W. I. Baggerman unpublished). Table 8 shows resuscitation factors of 1·1–7·2 that were obtained for 10 different quick-frozen vegetables. Similar results were obtained by us for Enterobacteriaceae and *E. coli* counts for frozen spinach

TABLE 8

*Plate counts of Enterobacteriaceae from quick frozen vegetables before and after resuscitation**

	Count of Enterobacteriaceae†		Resuscitation factor
	before resuscitation	after resuscitation	No. after resuscitation / No. before resuscitation
Onions	$4{\cdot}3 \times 10^7$	$4{\cdot}8 \times 10^7$	1·1
Vegetable mix 1	$1{\cdot}2 \times 10^6$	$2{\cdot}3 \times 10^6$	1·9
Chopped spinach	$1{\cdot}8 \times 10^5$	$4{\cdot}6 \times 10^5$	2·6
Sweet pepper	$1{\cdot}7 \times 10^4$	$3{\cdot}0 \times 10^4$	1·8
Leek	$0{\cdot}5 \times 10^4$	$2{\cdot}1 \times 10^4$	4·4
Vegetable mix 2	$1{\cdot}5 \times 10^3$	$8{\cdot}8 \times 10^3$	5·9
Curly kale	$3{\cdot}9 \times 10^3$	$8{\cdot}3 \times 10^3$	2·1
Endive	$1{\cdot}5 \times 10^2$	$5{\cdot}3 \times 10^2$	3·5
Butterbeans	$0{\cdot}5 \times 10^2$	$3{\cdot}6 \times 10^2$	7·2
Chervil	$0{\cdot}7 \times 10^2$	$1{\cdot}8 \times 10^2$	2·6

*Resuscitation for 60 min at 20°C in a 1:10 dilution in peptone–salt diluent.
†Per g of product.

in butter sauce. Most data available for Enterobacteriaceae or coliforms are based on enumeration without previous resuscitation, so it is in fact impossible to lay down specifications for frozen vegetables (which should include resuscitation procedures in the enumeration) because the essential basic data are missing. Preferable to an insistence on microbial specifications or guidelines would be the use of properly formulated codes for the Good Manufacturing Practice of frozen vegetables. Basic information for such exists already in a number of documents: the AFDOUS Frozen Food Code (of the Association of Food Drug Officials of the United States), the Current Good Manufacturing Practice (Sanitation) in manufacture, processing, packing or holding of Human Foods (Part 128 of the US Code of Federal Regulations), and the Recommended International Code of Practice and General Principles of Food Hygiene (of the FAO/WHO Codex Alimentarius Commission), all of which can be found in the excellent book, Prevention of Microbial and Parasitic Hazards Associated with Processed Foods (Anon. 1975).

The author is indebted to Miss M. Los for data of frozen vegetables and to Mr J. A. Drijber and Mrs L. R. Boerboom for skilful technical assistance.

10. References

ANDERSON, J. M. & BAIRD-PARKER, A. C. 1975 A rapid and direct plate method for enumerating *Escherichia coli* biotype I in food. *Journal of Applied Bacteriology* **39**, 111–117.

ANON. 1974*a* *Microorganisms in Foods* Vol. 2. Sampling for microbiological analysis: Principles and specific applications. Report of International Commission on Microbiological Specifications for Foods pp.105–109. Toronto: University of Toronto Press.

ANON. 1974*b* Frozen fruits and vegetables: general. *Israel Standard SI 877.* Tel Aviv: Standards Institution of Israel.

ANON. 1975 *Prevention of Microbial and Parasitic Hazards Associated with Processed Foods. A Guide for the Food Processor.* Washington DC: National Academy of Sciences.

ANON. 1977 *Frozen and Quick-Frozen Food.* Oxford: Pergamon Press.

ANON. 1979 Les spécifications applicables aux fruits et légumes surgelés. *La Surgelation* **168**, 55–67.

BERRY, J. A. 1933 Lactobacilli in frozen pack peas. *Science, N.Y.* **77**, 350–351.

BORNEFF, J. & WAGNER, M. 1967 Hygienisch-bakteriologische Untersuchungen an Tiefkühlgemüse. III. Mitteilung: Das Verhalten von Salmonellen an Tiefkühlgemüse. *Archiv für Hygiene und Bakteriologie* **151**, 83–90.

BROWN, E. B. 1933 Bacterial studies of defrosted peas, spinach, and Lima beans. *Journal of Home Economics* **25**, 887–892.

CAVETT, J. J., DRING, G. J. & KNIGHT, A. W. 1965 Bacterial spoilage of thawed frozen peas. *Journal of Applied Bacteriology* **28**, 241–251.

ERCOLANI, G. L. 1976 Bacteriological quality assessment of fresh marketed lettuce and fennel. *Applied and Environmental Microbiology* **31**, 847–852.

GELDREICH, E. E. & BORDNER, R. H. 1971 Fecal contamination of fruits and vegetables during cultivation and processing for market. A review. *Journal of Milk and Food Technology* **34**, 184–195.

HUCKER, G. J., BROOKS, R. F. & EMERY, A. J. 1952 The source of bacteria in processing and their significance in frozen vegetables. *Food Technology* **6**, 147–155.

JAMES, L. H. 1932 The microbiology of frozen foods. *The Fruit Products Journal and American Vinegar Industry* **12**, 110–113, 119.

JONES, A. H. & FERGUSON, W. E. 1956 The evaluation of frozen vegetables for quality. *Canadian Food Industries* **27**, 24, 27, 29, 31.

JONES, A. H. & LOCHHEAD, A. G. 1939 A study of micrococci surviving in frozen-pack vegetables and their enterotoxic properties. *Food Research* **4**, 203–216.

KÄFERSTEIN, F. K. 1976 The microflora of parsley. *Journal of Milk and Food Technology* **39**, 837–840.

KOZLOWSKI, A. V. 1977 Is it necessary to blanch all vegetables before freezing? In *Freezing, Frozen Storage and Freeze-drying of Biological Materials and Foodstuffs* pp.227–236. Paris: International Institute of Refrigeration.

LOCHHEAD, A. G. & JONES, A. H. 1936 Studies of numbers and types of micro-organisms in frozen vegetables and fruits. *Food Research* **1**, 29–39.

LOCHHEAD, A. G. & JONES, A. H. 1938 Types of bacteria surviving in frozen-pack vegetables. *Food Research* **3**, 299–306.

LUH, B. S. & O'NEAL, R. 1975 Quality Control. In *Commercial Vegetable Processing* ed. Luh, B. S. & Woodroof, J. G. pp.535–602. Westport, Connecticut: AVI Publishing Co.

MOHS, H. J. 1969 Hygiene bei Tiefgefrierkost. Bakteriologische Untersuchungen bei der Herstellung von tiefgefrorenen Gemüse. *Gordian* **69**, 374–377.

MUNDT, J. O. 1970 Lactic acid bacteria associated with raw plant food material. *Journal of Milk and Food Technology* **33**, 550–553.

MUNDT, J. O. 1976 Streptococci in dried and frozen foods. *Journal of Milk and Food Technology* **39**, 413–416.

MUNDT, J. O. & HAMMER, J. L. 1968 Lactobacilli on plants. *Applied Microbiology* **16**, 1326–1330.

PEDERSON, C. S. 1947 Significance of bacteria in frozen vegetables. *Food Research* **12**, 429-438.

RACCACH, M., JUVEN, B. & HENIS, Y. 1972 Variations in bacterial counts during the production of frozen green beans. *Journal of Food Technology* **7**, 417-421.

REEVES, M. P. 1973 Examination of frozen vegetables by two sample preparation procedures. *Journal of Food Science* **38**, 365-366.

SADOVSKI, A. Y. 1977 Technical note: acid sensitivity of freeze injured salmonellae in relation to their isolation from frozen vegetables by pre-enrichment procedure. *Journal of Food Technology* **12**, 85-91.

SMART, H. F. 1934 Microorganisms surviving the storage period of frozen-pack fruits and vegetables. *Phytopathology* **24**, 1319-1331.

SMART, H. F. 1937 Types and survival of some microorganisms in frozen-pack peas, beans, and sweet corn grown in the east. *Food Research* **2**, 515-528.

SMART, H. F. 1939 Microbiological studies on commercial packs of frozen fruits and vegetables. *Food Research* **4**, 293-298.

SMART, H. F. & BRUNSTETTER, B. C. 1937 Spinach and kale in frozen pack. I. Scalding tests. II. Microbiological studies. *Food Research* **2**, 151-163.

SPLITTSTOESSER, D. F. 1970 Predominant microorganisms on raw plant foods. *Journal of Milk and Food Technology* **33**, 500-505.

SPLITTSTOESSER, D. F. 1973 The microbiology of frozen vegetables. *Food Technology* **27**, 54, 56, 60.

SPLITTSTOESSER, D. F. & GADJO, I. 1966 The groups of microorganisms composing the 'total'-count population in frozen vegetables. *Journal of Food Science* **31**, 234-239.

SPLITTSTOESSER, D. F. & SEGEN, B. 1970 Examination of frozen vegetables for *Salmonella*. *Journal of Milk and Food Technology* **33**, 111-113.

SPLITTSTOESSER, D. F., WETTERGREEN, W. P. & PEDERSON, C. S. 1961*a* Control of microorganisms during preparation of vegetables for freezing. I. Green beans. *Food Technology* **15**, 329-331.

SPLITTSTOESSER, D. F., WETTERGREEN, W. P. & PEDERSON, C. S. 1961*b* Control of microorganisms during preparation of vegetables for freezing. II. Peas and corn. *Food Technology* **15**, 332-334.

SPLITTSTOESSER, D. F., HERVEY, G. E. & WETTERGREEN, W. P. 1965 Contamination of frozen vegetables by coagulase-positive Staphylococci. *Journal of Milk and Food Technology* **28**, 149-151.

TAMMINGA, S. K., BEUMER, R . R. & KAMPELMACHER, E. H. 1978 The hygienic quality of vegetables grown or imported into the Netherlands: a tentative survey. *Journal of Hygiene, Cambridge* **80**, 143-154.

TRESSLER, D. K. 1938 Bacteria, enzymes and vitamins—indices of quality in frozen vegetables. *Refrigeration Engineering* **36**, 319-321.

VAUGHN, R. H. & EDELWEISS, Z. Z. 1946 Control of microorganisms in frozen food plants. *Quick Frozen Foods* **9**, 76-78, 114.

WAGNER, M. & BORNEFF, J. 1967*a* Hygienisch-bakteriologische Untersuchungen an Tiefkühlgemüse. I. Mitteilung: Die Beschaffenheit der Handelsware. *Archiv für Hygiene und Bakteriologie* **151**, 64-74.

WAGNER, M. & BORNEFF, J. 1967*b* Hygienisch-bakteriologische Untersuchungen an Tiefkühlgemüse. II. Mitteilung: Ergebnis einer Betriebskontrolle. *Archiv für Hygiene und Bakteriologie* **151**, 75-82.

WHITE, A. & WHITE, H. R. 1962 Some aspects of the microbiology of frozen peas. *Journal of Applied Bacteriology* **25**, 62-71.

Toxic Bacterial Dusts Associated with Plants

MIRJA S. SALKINOJA-SALONEN AND ILKKA HELANDER

Department of General Microbiology, University of Helsinki, Helsinki, Finland

RAGNAR RYLANDER

Department of Environmental Hygiene, University of Gothenburg, Gothenburg, Sweden

Contents

1. Introduction

MANY human illnesses are known to be caused by inhaled vegetable dusts. The causal relationship of cotton dust to mill fever—a disease seen in young cotton workers—was described as early as in 1784 (see Pernis *et al.* 1961), and a host of other diseases—byssinosis, weaver's cough, hemp fever, grain fever and bagassosis—occurs among workers exposed to vegetable dusts. The purpose of this review is to reveal the inhalation toxicity of these dusts and to discuss the available evidence indicating that certain of these diseases and related clinical symptoms—Monday chest tightness, airway bronchoconstriction and chronic bronchitis—are caused by bacteria associated with the vegetable dusts, and more specifically, by endotoxins from certain of these bacteria.

2. Bacteria in Vegetable Dusts

Gram negative bacteria are present on most vegetable materials; they are found, for instance, on all parts of the cotton and flax plants. Cotton in the unopened boll is sterile but rapidly becomes contaminated once it opens in the field, and quantities ranging from 10^2–10^9 bacteria/g of material have been found in bale cotton from different countries (Rylander & Lundholm 1978). The bract is especially heavily contaminated by bacteria. Bagasse, hemp and mouldy hay also contain large amounts of Gram negative bacteria (Rylander *et al.* 1975).

When a cotton fibre is viewed with a scanning electron microscope, rod-shaped particles resembling bacteria can be seen attached to it (Fig. 1). No fungal hyphae or actinomycete-like structures are seen. The airborne flora in cotton mills is predominantly composed of Gram negative bacteria: these have been enumerated by Cinkotai *et al.* (1977), Cinkotai & Whitaker (1978), Fischer (1979), Haglind *et al.* (1981) and Bergström *et al.* (1980), and the number in cardrooms has been seen to range from 10^2 to 10^5 colony forming units (c.f.u.) per m^3. Dutkiewicz (1978) also demonstrated that high numbers of *Erwinia herbicola* (*Ent. agglomerans*) were present in the air of grain mills.

Both fermenting and non-fermenting Gram negative bacteria are found in cotton and flax (Table 1). *Enterobacter* spp., mainly *Ent. agglomerans*, constitute 40–100% of all Gram negative organisms found in cotton (Haglind *et al.* 1981). Of other fermenting species, *Klebsiella* spp. are often present. Among the non-fermenting Gram negative bacteria, strains of *Pseudomonas* are found, as well as *Acinetobacter* and *Agrobacterium*. Basically the same Gram negative flora is found on flax and in dust in flax mills (Symington 1980). *Escherichia coli* were not present.

3. Biological and Chemical Methods of Measuring the Amounts of Bacteria and Endotoxins in Vegetable Dusts

A. Endotoxins

Gram negative bacteria contain a substance with particular toxic properties —lipopolysaccharide (LPS, endotoxin). The LPS is located in the outer portion of the bacterial cell wall, where it makes up the lipid part of the

TABLE 1

Numbers of Gram negative bacteria on samples of various cotton products*

Type of sample	Alabama 1	Alabama 2	North Carolina 1	North Carolina 2	West Texas 1	West Texas 2	Western Australia 1	Western Australia 2	Western Australia 3	Western Australia 4
(a) Cotton plants										
Lint	3·8	3·1	4·2	5·0	6·2	6·7	4·7	5·3	5·6	6·1
Bract	3·3	5·2	6·5	7·7	7·0	7·6	5·8	6·3	8·8	7·1
Stem	6·2	4·8	7·0	7·0	6·7	6·5	3·5	4·1	6·9	6·1
Leaf	5·7	5·8	4·9	4·2	6·5	5·9	—	—	—	—

Type of sample	USA 1	USA 2	USA 3	USA 4	USA 5	USA 6	USA 7	USA 8	USA 9	USA 10	Russia 1	Russia 2	Australia	Turkey	Peru	Greece	Egypt
(b) Bale cotton	5·1	4·0	2·5	3·6	2·2	3·7	1·7	6·0	5·6	2·5	1·6	3·8	2·8	3·2	3·4	5·7	2·8

Type of sample	USA 1	USA 2	USA 3	USA 4	USA 5
(c) Cotton seed dust	5·9	3·9	4·0	5·6	3·9

*Expressed as $\log_{10}$ No. of bacteria/g of material.
Values for cotton of differing geographical origin should not be regarded as typical of that region, but merely indicative of the normal variation in values.
Data from Rylander & Lundholm (1978).

outer leaflet of the outer membrane. Chemically, endotoxins are macromolecules consisting of a carbohydrate chain anchored to a glucosamine disaccharide. The glucosamine residues are substituted by long-chain fatty acids. Endotoxins can be determined either by using their biological activity as an index, or by chemical methods.

B. Biological methods

Cavagna *et al.* (1969) assayed endotoxin in cotton dust by induction of necrosis on epinephrine-injected skin sites of rabbits. They reported 1·4–1·6 mg of an endotoxin-like substance/g of cardroom dust, with commercial endotoxin as the reference.

The *Limulus* lysate method (Wildfeuer *et al.* 1974) can also be used. Cinkotai *et al.* (1977) determined the concentration of air-borne endotoxin in cotton mills using this method. Values ranging from 0·2 to 1·6 $\mu g/m^3$ were found in cardroom air. Similar concentrations were later reported by Fischer (1979). Bergström *et al.* (1980) reported values ranging from 0·3 to 2·5 $\mu g/m^3$ in an experimental cardroom where ordinary bale cotton was carded under conditions resembling those in cotton mills. Symington (1980) found levels of up to 2 $\mu g/m^3$ in the air of flax mills using the *Limulus* method.

C. Chemical methods

Endotoxins can also be determined by measuring the fatty acids with gas chromatography. This method can be considered specific, as the spectrum of fatty acids in bacterial endotoxins is quite different from that in higher organisms, including plants (Wilkinson 1977).

Fatty acid analysis of vegetable dust offers a means of assessing total bacterial contamination (both viable and dead). Table 2 shows the results of such analyses. Myristic, palmitoleic, stearic and *cis*-vaccenic acid (distinguishable by gas–liquid chromatography from the other ^{18}C acids which occur in plants) constituted over 90% of all fatty acid found in the cotton dust samples that have been analysed in these laboratories. This spectrum of fatty acids is typical of lipids in Gram negative bacteria (Goldfine 1972). It can be estimated from the results that *ca.* 3–10% of the dry weight of dust, and 0·4% of cotton, might consist of bacterial biomass. If a correction is made for recovery, which seems to be about 50% in the procedure used here (of 10 mg/g of added agrobacteria, lipids corresponding to 4·5 mg were recovered), these figures could be multiplied by a factor of 2. This estimate (*ca.* 1% of bacterial biomass in cotton No. 2) is well in accordance with the impression of bacterial mass obtained in Fig. 1.

TABLE 2

The amount of bacterial biomass in cotton dust and cotton: estimation on basis of fatty acid analysis

Sample	Wt (μg) of fatty acid*/g material						Equivalent to dry bacteria† (mg/g material)
	14:0	16:1	16:0	18:1	18:0	total	
Crescent 770418C	28	18	590	426	97	1159	35
Moss 770330	86	82	1740	1091	295	3294	98
Dry cotton fibre (2)	4·6	3·4	55	29	37	129	3·9
Dry cotton fibre (2) & 1% of dry bacteria‡	5·8	7·8	78	159	38	289	8·6

*Lipid was extracted from the material by a mixture of chloroform–methanol 3:1, followed by separation of phases (Folch partition). Fatty acid was liberated from the lipid by hydrolysis in 4N KOH, 5 h at 100° C, derivatized and analysed as described in Salkinoja-Salonen and Boeck (1978).

†Assuming that 3% of the bacterial biomass is accounted for by the fatty acids.

‡*Agrobacterium tumefaciens* TT111, lyophilized.

4. Inhalation Effects of Gram Negative Bacteria and Endotoxins

A. The importance of responses to endotoxin

The adverse effects of Gram negative bacterial septicaemia and parenterally administered endotoxins are well known (Wolff 1973). Many of the adverse effects of Gram negative bacteria on animals and humans—ranging from fever, complement activation, macrophage activation and mitogenicity to lethal effects—are attributed to the endotoxin (see Galanos *et al.* 1977). Differences in response to endotoxin are noted between species, and mice, rats and hamsters have been found to be relatively insensitive to endotoxin, whereas guineapigs, and particularly man, are sensitive (Kuida *et al.* 1961).

Considerably less information is available on the effects of inhaled endotoxin, the relevant exposure route relating to vegetable dusts, than on those of other routes of exposure. In the following text, the particular effects of inhalation will be reviewed, based upon experience from toxicological and epidemiological studies.

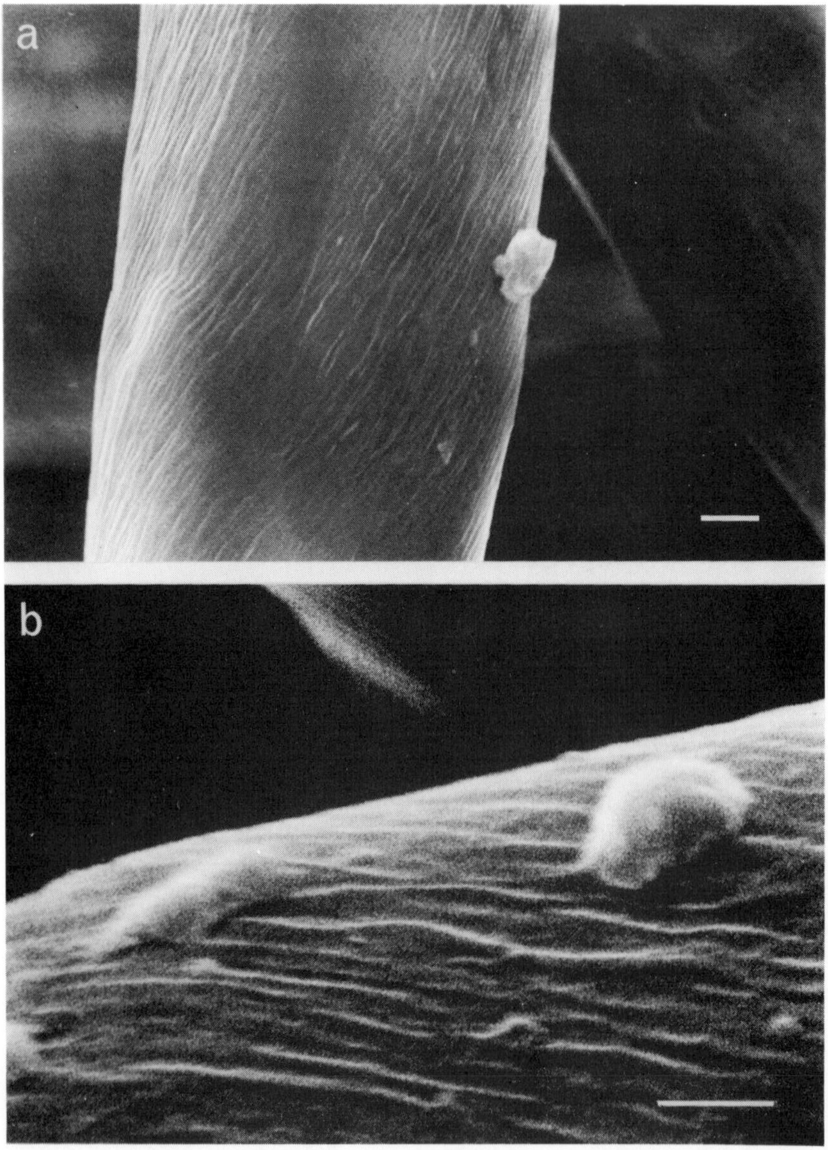

Fig. 1. Scanning electron micrograph of cotton fibres demonstrating structures similar to bacteria on the surfaces. (*a*) Bar = 2 μm, (*b*) Bar = 1 μm.

B. Inflammation

Exposure to cotton dust or extracts thereof will initiate an inflammatory response in the airways. Cavagna *et al.* (1969) exposed rabbits to an aerosol of cotton extract daily from Monday to Friday for 20 weeks. The animals showed pulmonary changes in terms of exfoliation of bronchial cells, intra-alveolar septal thickening, peribronchial lymphocyte infiltration and an increased excretion of protein-containing fluid from the respiratory epithelium. These changes were observed in 4–6 animals out of 10. No such changes were seen in the control group receiving saline aerosol only. Similar changes were found in rabbits that had inhaled aerosol prepared from commercial *E. coli* endotoxin. A dose of 20 μg of endotoxin in 2 ml of saline produced the same pulmonary changes as 2 mg of cotton dust in 2 ml given daily. Rylander & Nordstrand (1974) used guineapigs as their animal model. They found that exposure to an aerosol of cotton bract dust mobilized polymorphonuclear leukocytes and, to a lesser degree, macrophages in the airways.

This ability of Gram negative bacteria and endotoxin to induce into the airways a neutrophil invasion, indicative of early inflammation, was later demonstrated in other experiments (Walker *et al.* 1975, Rylander & Snella 1976, Hudson *et al.* 1977). The reaction is dose-related and for various species of animals it differs in magnitude.

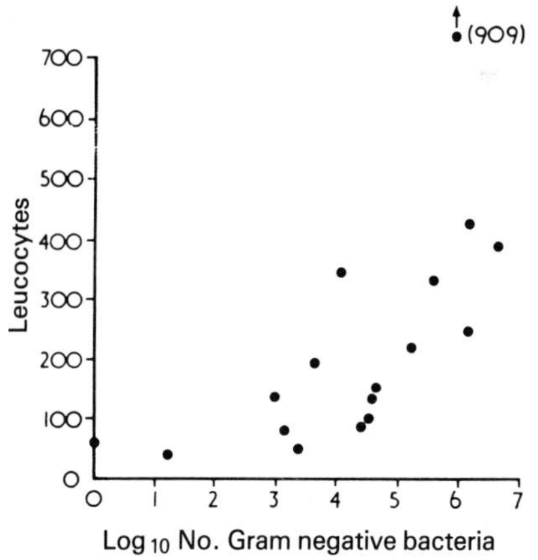

Fig. 2. Relationship between neutrophil invasion in the airways of guineapigs and the number of Gram negative bacteria in different cotton dust extracts (Rylander & Snella 1976).

Figure 2 shows the relationship between the content of Gram negative bacteria in the dust and the extent of neutrophil invasion into the airways of guineapigs (Rylander & Snella 1976). Table 3 shows that the

TABLE 3
Lowest concentrations of dust giving a positive Limulus *reaction*

Preparation	Dry wt (μg) dust per ml	Leucocytes*
Bract		
73-4	41·7	54
73-6	5·2	190
73-9	5·2	344
73-10	5·2	141
73-11	10·4	135
73-121	2·1	388
73-13	10·4	83

*Number of leucocytes ($\times 10^5$) in airways of guineapigs exposed to aerosolized extracts of the different dusts.
Data from Rylander & Snella (1976).

capacity to mobilize neutrophils into the airways of guineapigs is related to the content of endotoxin of that bract sample. The airway neutrophil response becomes progressively smaller in guineapigs exposed subacutely, and there is reason to believe that this is dependent upon the development of a local immune response (Rylander *et al.* 1980). The neutrophil invasion into the airway epithelium has also been demonstrated in humans among workers in a cotton mill (Merchant *et al.* 1975) and cotton mill workers exposed in an experimental cardroom (Bergström *et al.* 1980). Even those cotton mill workers and students with no previous exposure to cotton dust have also been shown to differ in their neutrophil response after exposure in an experimental cardroom (Rylander 1981).

The degree of neutrophil migration differs between different strains of bacteria found in cotton. Table 4 shows that some bacteria (*Enterobacter*, *Klebsiella*, *Pseudomonas*) have a large capacity for mobilizing neutrophils, whereas other bacteria have a lower capacity (*Agrobacterium*, *Bacillus*). Agrobacteria, although Gram negative and possessing endotoxin, did not mobilize the neutrophils. In further experiments (Helander *et al.* unpublished), it has been seen that extracts made of *Xanthomonas sinensis* and the blue-green algae (cyanobacteria) *Phormidium uncinatum*, *Ph. africanum*, *Anabaena variabilis* and *Synechococcus* sp. were inactive as well. Suspensions of *Citrobacter*, however, were extremely active.

Table 5 (Helander *et al.* 1980) shows the number of free lung cells that migrate into the lungs of guineapigs after exposure to aerosols of pure endotoxins from *Ent. agglomerans*, *Ps. putida*, *K. oxytoca* and *Agrobacterium* sp. (all isolated from cotton, as well as from *E. coli* (commercial preparation) and *Xanthomonas sinensis*. It is seen that the purified endotoxin (LPS) alone induced the migration of neutrophils and

TABLE 4

Number of macrophages and leucocytes in airways of guineapigs exposed to aerosols of different bacteria from cotton plants

Bacteria in aerosol	Number of animals	Macrophages (mean No. × 5 × 10⁴/lung)	Leucocytes (mean No. × 5 × 10⁴/lung)
Control	25	145 (49)*	65 (40)
Enterbacter cloacae	10	452† (235)	662† (422)
Enterobacter agglomerans	10	402† (193)	527† (204)
Agrobacterium spp.	10	149 (41)	86 (32)
Klebsiella oxytoca	10	407† (209)	663† (306)
Pseudomonas syringae	10	417† (186)	596† (296)
Bacillus spp.	9	153 (34)	91 (32)

*Standard deviation in parentheses.
†Statistically significant difference from control value (Student's t test), $P < 0{\cdot}001$.
Data from Rylander & Lundholm (1978).

TABLE 5

*Mean numbers × 10^6 of free lung cells 24 h after a 40-min exposure to an aerosol suspension of 5 µg/ml LPS in sterile water**

LPS	Total cells	Macrophages	Lymphocytes	Neutrophils	Eosinophils
Enterobacter agglomerans	53·4 ± 16·0†	25·7 ± 8·1†	3·6 ± 1·6 †	20·1 ± 9·0†	4·1 ± 2·3
Pseudomonas putida	43·9 ± 19·5†	25·8 ± 11·4†	2·3 ± 1·0	12·1 ± 5·4†	3·6 ± 2·9
Klebsiella oxytoca	83·4 ± 22·3†	35·8 ± 12·6†	5·1 ± 1·0†	35·6 ± 20·2†	6·9 ± 2·8†
Agrobacterium sp.	15·5 ± 7·4	9·2 ± 4·3	1·2 ± 0·5	2·9 ± 2·5	2·2 ± 1·2
Escherichia coli	23·1 ± 7·0†	15·2 ± 4·5	1·1 ± 0·7	4·5 ± 2·1†	2·2 ± 1·3
Xanthomonas sinensis	17·8 ± 6·1	13·0 ± 4·1	1·4 ± 0·7	1·5 ± 0·9	1·9 ± 1·0
Control	12·6 ± 4·4	9·7 ± 4·0	1·0 ± 0·5	0·4 ± 0·1	1·6 ± 0·7

*Mean ± standard deviation for 10 animals.
†$P < 0{\cdot}001$ (t test, Welch modification).

macrophages in the same way as did the whole bacterial cells. Endotoxins from *Agrobacterium* sp. and *Xanthomonas* were much less potent inducers of neutrophil migration than the others. The differences are probably related to the chemical structure of the respective endotoxins. The endotoxins were analysed and found to be pure (< 1% contamination by protein or nucleic acid) (Helander *et al.* 1980). The composition of endotoxins of the strains isolated from cotton were found to be similar to that reported in the literature for these genera (for review see Galanos *et al.*

1977; Wilkinson 1977). *Agrobacterium* and *Xanthomonas* lipopolysaccharides contain glucose, mannose, glucosamine and KDO (2-keto-3-deoxy-octonic acid), but lack galactosamine and heptose (Volk 1968; Salkinoja-Salonen & Boeck 1978), components that are present in the other lipopolysaccharides shown in Table 5. Absence of heptose does not alone account for the lack of neutrophil recruiting activity, however, for when LPS containing heptose but not KDO was tested, it was also found to be inactive (Helander *et al.* 1980).

The fatty acid composition of the endotoxins with pulmonary activity differs from those that are void of activity: *X. sinensis* contains iso-branched hydroxy fatty acids of fewer than 13 carbon atoms/molecule and non-hydroxy fatty acids of 10 and 11 carbon atoms (Rietschel *et al.* 1975). The agrobacterial LPS contains little non-hydroxy fatty acid (< 2% w/w), the hydroxy fatty acids being 3-hydroxy myristic and palmitic acids (Salkinoja-Salonen & Boeck 1978). The biological properties of lipids depend very much on their fatty acid composition. Chain length and hydrophobic properties of the fatty acids determine the fluidity of the membrane (for review see Singer 1974). All the strains that had little effect on neutrophil migration into the lung lacked (or contained very little of) *n*-chain fatty acid of 12–16 carbon atoms/molecule in their LPS (for the fatty acids of the blue-green algae LPS see Keleti *et al.* 1979).

These results thus agree with the theory (Bradley 1979) that fatty acid is an important functional component of the LPS toxophore. Furthermore, the results indicate that endotoxin is responsible for the pulmonary effects caused by inhaled whole bacteria, and that the severity of these effects depends on the structure of the endotoxin, probably the Lipid A portion. Other activities of endotoxin (pyrogenicity, lethal toxicity, complement reactivation) are also known to depend on Lipid A structure (Lüderitz *et al.* 1978).

C. Pulmonary function

Exposure to cotton dust or extracts will cause bronchoconstriction in animals and humans. Although only sparse data are available, the same reaction is probably caused by Gram negative bacteria and endotoxin. Cavagna *et al.* (1969) exposed normal human subjects and subjects with mild chronic bronchitis to an aerosol containing 40–80 μg of purified *E. coli* endotoxin in 2 ml of saline. In two out of eight normal subjects, a significant reduction of $FEV_{1\cdot 0}$ (Forced Expiratory Volume) was seen after inhalation of 80 μg of endotoxin. Exposure to considerably lower doses of endotoxin was performed in a study of shower taker's disease (Muittari *et al.* 1980*a*,*b*). Using tap water containing about 1 μg of endotoxin/ml, a one millilitre inhalation challenge brought about a decreased lung diffusion capacity 4–6 h after the exposure. No studies of $FEV_{1\cdot 0}$ were performed in this experiment.

The airway constriction is probably mediated through histamine release. Endotoxins cause the release of histamine (Hinshaw *et al.* 1961, Davis *et al.* 1963), but the mechanism for this effect is not clear. Potential mechanisms are: an effect of macrophages on platelets, disruption of neutrophils with release of intracellular histamine or an effect on mast cells.

D. Epidemiological studies

An association between bacteria on cotton plants and pulmonary illnesses was first suggested by Neal *et al.* (1942) in a report on mattress makers working with a low grade, stained cotton. The symptoms were similar to those seen in 'mill fever' among cotton mill workers. The authors examined samples of the cotton used, and exposed animals and humans to the dust or extracts thereof. They concluded that the illness was caused by the inhalation of Gram negative micro-organisms, or their products, which were present on stained cotton and its dust.

Cinkotai & Whitaker (1978) studied the extent of byssinosis symptoms, using a questionnaire, and the concentrations of airborne micro-organisms and endotoxins in 21 cotton spinning and cotton waste mills in Lancashire, UK. Dust concentration (total dust less fly), airborne bacteria and endotoxin were measured in the various work places. No correlation was seen between the dust concentration and byssinotic symptoms; in fact, only few byssinotic workers were found in some of the very dusty mills, while other, cleaner, mills showed a high prevalence. The correlation between prevalence of symptoms and the amount of airborne Gram negative bacteria, however, was highly significant. No relationship was seen between byssinosis prevalence and the concentration of airborne fungi.

In a study on cotton mill workers in Sweden, Haglind *et al.* (1981) also used a questionnaire to relate the extent of byssinosis to the amount of airborne dust and the number of viable Gram negative bacteria. The former correlated with byssinosis extent in all mills except a cardroom for medical cotton. When this site was included in the material, the number of Gram negative bacteria gave a better correlation with the extent of byssinosis among the workers than did the dust levels.

Determinations of lung function are used for objective measurements of effects induced by vegetable dust exposure. Rylander *et al.* (1979) measured the average decrease in $FEV_{1\cdot 0}$ over the Monday shift period, dust levels (measured by vertical elutriator) and the number of Gram negative bacteria in the bale cotton being processed in 23 cotton mills in the USA. Figure 3 shows the relationship between the $\Delta FEV_{1\cdot 0}$ values and the number of Gram negative c.f.u. in the bale cotton (r_{xy} = 0·78). The dose response relationship suggests a threshold value of *ca.* 10^3 Gram negative bacteria/g of bale cotton.

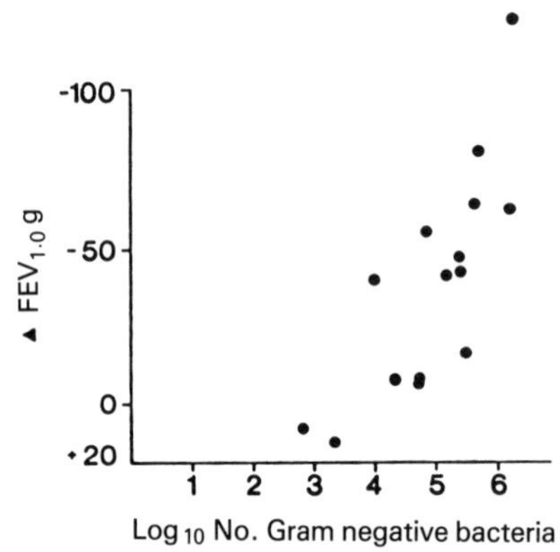

Fig. 3. Relationship between decrease in $FEV_{1\cdot 0}$ over Monday shift working period and Gram negative bacteria in bale cotton (Rylander *et al.* 1979).

E. Case histories

Microbiological work with cotton and flax focussed attention on two persons with similar histories of development of pulmonary symptoms. A 48-year-old generally healthy woman complained of attacks of high fever (39–40°C), aches in the extremities and a feeling of malaise occurring shortly after she had started to work in a greenhouse. The symptoms disappeared within a day (except for a dry cough) and never reappeared in any other than a greenhouse environment with geranium plants. The number of viable airborne bacteria in the greenhouse was measured and it was found that the normal level of *ca.* 60–100 c.f.u./m^3 increased to a level of 410–1290 c.f.u./m^3 after the geraniums had been handled for a few minutes. Determinations were made of the number of c.f.u. on Drigalsky agar from all the different plants in the greenhouse, and it was found that the geraniums carried up to 10^7 Gram negative bacteria/g of fresh leaves (extracted with saponin solution).

Another case involved a man of 46 years of age, who complained of chills, dyspnoea, slight temperature increase (37°C) and a feeling of malaise on Mondays or other occasions when he washed bean sprouts from a commercial growing process. He had been treated by a physician who found a 60% decrease in transfer ratio, transient increase of white cell count in blood (up to $16{\cdot}9 \times 10^6$/ml), but no changes in radiographic appearance. The patient did not experience symptoms if he used a gas mask during that particular stage of his work. Plate counts were made on tryptose–glucose–yeast extract agar, malt agar and potato–dextrose agar plates of the various bean washes and air of the working environment. *Enterobacter agglomerans* was found in the bean sprout wash water and a large number of airborne bacteria was found while the seeds were being washed under a warm shower.

In neither case was there any significant level of contamination of the air by fungi ($<10^2$/ml as c.f.u. on PDA or malt agar). Actinomycete colonies were seen only sporadically.

5. Summary

This review of the inhalation toxicity properties of Gram negative bacteria and endotoxins demonstrates that several of the clinical symptoms found among workers exposed to vegetable dusts, such as mill fever, Monday chest tightness, airway bronchoconstriction and chronic bronchitis, can be explained by this exposure. Although the evidence is not complete, an overall evaluation suggests a theory of causal relationship.

We would like to thank Professor Y. Ovodov (Pacific Institute of Bio-organic Chemistry, Vladivostok, USSR), Dr E. T. Rietschel (Max Planck Institut für Immunbiologie, Freiburg, BDR) and Dr J. Weckesser (University of Freiburg, Biologie III, BDR) for their gifts of cells and purified lipopolysaccharide of the *Phormidium*, *X. sinensis*, *A. variabilis* and *Synechococcus*, respectively.

Dr Kari Lounatmaa (Department of Electron Microscopy, University of Helsinki, Finland) has prepared the scanning electron micrographs of cotton fibre (Fig. 1), which we gratefully acknowledge. Cotton Incorporated, the Swedish Work Environment fund, the Academy of Finland and Suomen Kulttuurirahasto have given financial support.

6. References

BERGSTRÖM, R., HAGLIND, P. & RYLANDER, R. 1980 Experience from human exposures in an experimental cardroom. In *Proceedings of the Fourth Session on Cotton Dust Research, Beltwide Cotton Production Research Conference, St. Louis, Missouri* ed. Wakelyn, P. J. pp.22–25.

BRADLEY, S. G. 1979 Cellular and molecular mechanisms of endotoxin action. *Annual Review of Microbiology* **33**, 67–94.

CAVAGNA, G., FOA, V. & VIGLIANI, E. C. 1969 Effects in man and rabbits of inhalation of cotton dust or extracts and purified endotoxin. *British Journal of Industrial Medicine* **26**, 314–321.

CINKOTAI, F. F. & WHITAKER, C. J. 1978 Airborne bacteria and the prevalence of byssinotic symptoms in 21 cotton spinning mills in Lancashire. *Annals of Occupational Hygiene* **21**, 239–250.

CINKOTAI, F. F., LOCKWOOD, M. G. & RYLANDER, R. 1977 Airborne bacteria and the prevalence of byssinotic symptoms in cotton mills. *American Industrial Hygiene Association Journal* **38**, 554–559.

DAVIS, R. B., BAILEY, W. L. & HANSON, N. P. 1963 Modification of serotonin and histamine release after *E. coli* endotoxin administration. *American Journal of Physiology* **205**, 560–566.

DUTKIEWICZ, J. 1978 Exposure to dust borne bacteria in agriculture. *Archives of Environmental Health* **33**, 250–269.

FISCHER, J. J. 1979 The microbial composition of cotton dusts, raw cotton lint samples and the air of carding areas in mills. In *Proceedings of the Third Special Session on Cotton Dust Research, Beltwide Cotton Production Research Conference, Phoenix, Arizona* ed. Wakelyn, P. J. pp.8–10.

GALANOS, C., LÜDERITZ, O., RIETSCHEL, E. T. & WESTPHAL, O. 1977 Newer aspects of the chemistry and biology of bacterial lipopolysaccharides, with special reference to their lipid A component. In *International Review of Biochemistry, Biochemistry of Lipids II* Vol. 14, ed. Goodwin, T. W. pp.239–337. Baltimore: University Park Press.

GOLDFINE, H. 1972 Comparative aspects of bacterial lipids. *Advances in Microbial Physiology* **8**, 1–51.

HAGLIND, P., LUNDHOLM, M. & RYLANDER, R. 1981 Byssinosis prevalence in Swedish cotton mills. *British Journal of Industrial Medicine* **38**, 138–143.

HELANDER, I., SALKINOJA-SALONEN, M. & RYLANDER, R. 1980 Chemical structure and inhalation toxicity of lipopolysaccharides from bacteria on cotton. *Infection and Immunity* **29**, 859–862.

HINSHAW, L. B., JORDAN, M. M. & VICK, J. A. 1961 Mechanism of histamine release in endotoxin shock. *American Journal of Physiology* **200**, 987–989.

HUDSON, A., KILBURN, K., HALPRIN, G. & MCKENZIE, W. 1977 Granulocyte recruitment to airway exposed to endotoxin aerosol. *American Review of Respiratory Diseases* **115**, 89–95.

KELETI, G., SYKORA, J. L., LIPPY, E. C. & SHAPIRO, M. A. 1979 Composition and biological properties of lipopolysaccharides isolated from *Schizothrix calcicola* (Ag.) Gomont (cyanobacteria). *Applied and Environmental Microbiology* **38**, 471–477.

KUIDA, H., GILBERT, R. P., HINSHAW, L. B., BRUNSON, J. G. & VISSCHER, M. B. 1961 Species differences in effect of Gram-negative endotoxin on circulation. *American Journal of Physiology* **200**, 1197–1202.

LÜDERITZ, O., GALANOS, C., LEHMANN, V., MAYER, H., RIETSCHEL, E. TH. & WECKESSER, J. 1978 Chemical structure and biological activities of lipid A's from various bacterial families. *Naturwissenschaften* **65**, 578–585.

MERCHANT, J. A., HALPRIN, G. G., HUDSON, A. R., KILBURN, K. H., MCKENZIE, W. N., HUNT, D. J. & BERMAZOHN, P. 1975 Responses to cotton dust. *Archives of Environmental Health* **30**, 222–229.

MUITTARI, A., KUUSISTO, P., VIRTANEN, P., SOVIJÄRVI, A., GRÖNROOS, P., HARMOINEN, A., ANTILA, P. & KELLOMÄKI, L. 1980*a* An epidemic of extrinsic allergic alveolitis caused by tap water. *Clinical Allergy* **10**, 77–90.

MUITTARI, A., RYLANDER, R. & SALKINOJA-SALONEN, M. 1980*b* Endotoxin and bath-water fever. *Lancet* **ii**, 89.

NEAL, P. A., SCHNEITER, R. & CAMINITA, B. H. 1942 Report on acute illness among rural mattress makers using lowgrade stained cotton. *Journal of the American Medical Association* **119**, 1074–1082.

PERNIS, B., VIGLIANI, E. C., CAVAGNA, C. & FINULLI, M. 1961 The role of bacterial endotoxins in occupational diseases caused by inhaling vegetable dusts. *British Journal of Industrial Medicine* **18**, 120–129.

RIETCHEL, E. TH., LÜDERITZ, O. & VOLK, W. A. 1975 Nature, type of linkage, and absolute configuration of (hydroxy) fatty acids in lipopolysaccharides from *Xanthomonas sinensis* and related strains. *Journal of Bacteriology* **122**, 1180–1188.

RYLANDER, R. 1981 Bacterial toxins and ethiology of byssinosis. *Chest* **79**, 345–385.

RYLANDER, R. & LUNDHOLM, M. 1978 Bacterial contamination of cotton and cotton dust and effects on the lung. *British Journal of Industrial Medicine* **35**, 204–207.

RYLANDER, R. & NORDSTRAND, A. 1974 Pulmonary cell reactions after exposure to cotton dust extract. *British Journal of Industrial Medicine* **31**, 220–223.

RYLANDER, R. & SNELLA, M-C. 1976 Acute inhalation toxicity of cotton plant dusts. *British Journal of Industrial Medicine* **33**, 175–180.

RYLANDER, R., NORDSTRAND, A. & SNELLA, M-C. 1975 Bacterial contamination of organic dusts. Effects on pulmonary cell reactions. *Archives of Environmental Health* **30**, 137–140.

RYLANDER, R., IMBUS, H. R. & SUH, M. W. 1979 Bacterial contamination of cotton as an indicator of respiratory effects among card room workers. *British Journal of Industrial Medicine* **36**, 299–304.

RYLANDER, R., MATTSBY, I. & SNELLA, M-C. 1980 Airway immune response after exposure to inhaled endotoxin. *Bulletin Européen de Physiopathologie Respiratoire* **16**, 501–509.

SALKINOJA-SALONEN, M. & BOECK, R. 1978 Characterization of lipopolysaccharides isolated from *Agrobacterium tumefaciens*. *Journal of General Microbiology* **105**, 119–125.

SINGER, M. 1974 The molecular organization of membranes. *Annual Review of Biochemistry* **43**, 805–833.

SYMINGTON, I. S. 1980 Respiratory symptoms in flax workers in Scotland. *Chest* in press.

VOLK, W. A. 1968 Quantitative assay of polysaccharide components obtained from cell wall lipopolysaccharides of *Xanthomonas* species. *Journal of Bacteriology* **95**, 980–982.

WALKER, R. F., EIDSON, G. & HATCHER, J. D. 1975 Influence of cotton dust inhalation on free lung cells in rats and guinea pigs. *Laboratory Investigation* **33**, 28–32.

WILDFEUER, A., NEYMER, B., SCHLEIFER, K. H. & HAFERKAMP, O. 1974 Investigations on the specificity of the limulus test for the detection of endotoxin. *Applied Microbiology* **28**, 867–871.

WILKINSON, S. G. 1977 Composition and structure of bacterial lipopolysaccharides. In *Surface Carbohydrates of the Prokaryotic Cell* ed. Sutherland, I. pp.97–175. London & New York: Academic Press.

WOLFF, S. M. 1973 Biological effects of bacterial endotoxins in man. In *Bacterial Lipopolysaccharides* ed. Kass, E. H. & Wolff, S. M. pp.251–256. University of Chicago Press.

Light Microscope Techniques for the Microbiological Examination of Plant Materials

A. M. PATON

Division of Agricultural Bacteriology, School of Agriculture, University of Aberdeen, Aberdeen, UK

Contents

1. Introduction

IT IS A common practice to consider the existence of the higher plants in isolation from other living macro- or micro-organisms. A more enlightened approach admits a relationship whereby the plants gain or lose in a form of competition. Acceptance of this outlook suggests, therefore, that by modifying the nature of the fauna and flora on and around plants we can favour growth of the plants. This is, to some extent, what plant husbandry is all about. Despite this we tend to under-rate the role of the epiphytic biomass and, by reason of our present ignorance, we fail to take full advantage of one of the most efficient natural complexes that has evolved.

It is not the purpose of this paper to consider the nature and functions of the organisms intimately associated with plants. Neither is it appropriate here to speculate on the future developments of such knowledge. It suffices to emphasize that the relationships among micro-organisms and those between micro-organisms and plants are likely to gain in importance not only in terms of plant health but also in the revelation of new concepts in multicomponent microbiology. New understanding of this kind, however, awaits the development of new technology or better use of older technical skills for it is a poor scientist who ignores the well-tried and tested procedures of the past.

There is a present tendency among many microbiologists to consider light microscopy as a superficial or minor skill, or at best, useful for simple determinations of a morphological nature. The world of micro-organisms must be revealed visually before the application of aids to identification and function. The examination of organisms in isolation *in vitro* without adequate reference to their ecology can be grossly misleading. Under the microscope the natural relationships are revealed to provide unlimited scope for the imaginative observer.

The plant microscopist has more to gain than most from acute observation by microscopy. The particular technology associated with the examination of plant surfaces has evolved over many years, stimulated from time to time by the needs of the pathologist and the constant interests in free-living and symbiotic nitrogen-fixing systems. In recent years the study of the epiphytic flora has gained in importance (Dickinson & Preece 1976) and it is therefore appropriate that the methods commonly employed should be reconsidered and suggestions made for their improvement. While omitting details of common knowledge and experience the following remarks are intended to extend our ability to view the plant surface and its superficial residents. Beyond these words there is ample opportunity for continued technical improvement.

The methods suggested in this paper comprise only a proportion of those available for adaptation from varied microbiological interests. They are deliberately dealt with in outline only, as each can be readily varied according to the individual requirements of the user. This area of microscopy is in a state of rapid development and those workers wishing to exploit the techniques for studies in plants may well lead that progress.

2. The Examination of Plant Surfaces

When methods are devised for the examination of plant surfaces they should be applicable not only to the phylloplane and rhizoplane but also to the external and internal surfaces of the seed testa, a much neglected but important habitat for micro-organisms. Whatever surface is entailed, the observation of fungi and bacteria *in situ* is inevitably made difficult by the surface irregularity, by the problems associated with the differentiation of the organisms from the plant, and by the numerical losses of cells occurring during manipulation, staining, washing and mounting.

The irregular nature of the plant surface is a disturbing visual feature in most preparations—the more so at high magnifications when the decreased depth of focus allows acute sight of only a portion of the microscope field. The observer can, to some extent, ignore this by continual focussing but a photograph cannot offer such a compensation. It is thus impossible, in most cases, to photograph with accuracy the image conveyed to the eye. All the methods involving the direct examination of the surfaces suffer from this disadvantage. Indirect examinations using impression films on materials such as cellulose acetate or clear adhesive tapes (Butler & Mann 1959; Dickinson 1974) avoid this difficulty to some extent but, at best, only a proportion of the epiphytes adhere to the films, thus providing a relatively poor representation of the natural situation. For many purposes, however, a representation of this kind is acceptable.

A. The differentiation of micro-organisms from plant materials by transmitted light microscopy

(i) *Pre-staining procedures*

The differentiation of organisms from plant material is less difficult and many stains are available which have a stronger affinity for the epiphytic flora than for the intact plant surface. This differentiation is greatly improved if the tissue background is cleared by solvents or by bleaching, processes which are obligatory, in any case, for transmitted light microscopy (Bevege 1968; Fowler & Owen 1971; O'Brien & Von Teichman 1974). Bleaching by hypochlorite solutions or by chlorine gas is rapid and effective (Janes 1962) but its use is mainly confined to situations where the detailed structure of the organism or the plant tissue is unimportant. The use of acridine orange and immunofluorescent staining are greatly impaired by bleaching processes. Clearing with ethanol, or with a mixture of formalin, acetic acid and ethanol, is acceptable for the immunofluorescent technique, for most of the traditional stains and for many fluorochromes. All clearing processes, of course, preclude the possibility of determining viability.

It is evident that the true epiphytic flora is sometimes more difficult to dislodge from plant surfaces than might be supposed (Rovira *et al.* 1974), but, nevertheless, when critical examinations or counts are required care must be taken to minimize losses from the surfaces. Clearing with ethanol or with other fixatives containing ethanol commonly allows an acceptable retention of cells even during subsequent aqueous procedures. The degree of retention afforded by a particular method may be determined by a membrane filtration examination of the used fluids at the different stages of preparation.

Thin fragments of plant tissue such as, for example, most leaves and rootlets, present little difficulty during the clearing process, but thicker structures require unduly long periods of treatment. Even if clearing has been effective the preparation is often unsuitable for examination under the microscope due to the distorting effect and opacity of the underlying tissue.

The epidermal layers of plant tissue can often be removed by pectinase treatment. The tissues are fixed with ethanol, thoroughly but gently washed and immersed in a pectinase-buffer solution for several hours at 25–27°C. The epidermal layer can then be separated from the rest of the plant tissue, stained and mounted in the usual manner. The pectinase appears to have little, if any, effect on the epidermal structures, and the adherence of micro-organisms is maintained. Such a process is, however, time consuming and should only be considered in special cases.

(ii) *Stains and staining*

Among the numerous stains which have been recommended in the literature (Rovira 1956; Preece 1959; Ruinen 1961; Dickinson 1967; Fowler & Owen 1971), only a few such as aniline blue, acid fuchsin or erythrosine with lactophenol, are of common useage. Periodic acid with Schiff's reagent is also the choice of many plant biologists. Experience, as usual, assists success in the satisfactory use of these traditional stains. The production of high quality stained plant preparations is often more of an art than a science, but a poor choice or use of fixative and overstaining are frequent causes of ineffective differentiation. Efficient use of the microscope is rarely encountered nowadays—the instrument is too often considered as a simple tool and its finer adjustments ignored or misunderstood. Details such as the thickness of the coverglass, diaphragm and condenser adjustment and the use of appropriate colour filters are all as important as a well prepared specimen. It is appropriate at this point to draw due attention to the carcinogenic nature, proven or alleged, of the components of some common fixatives and stains. In the absence of safe alternatives, their use is likely to continue albeit with greater care than in the past.

B. Fluorescence microscopy

The traditional methods of clearing, staining and observation with ordinary transmitted light microscopy undoubtedly have a permanent place in the laboratory for they have stood the test of time and have proved to be invaluable as reliable means of studying organisms on and in plant tissues. However, the advent of the fluorescence microscope, particularly instruments utilizing incident or epifluorescent illumination has provided a most useful addition to the technology available to plant scientists. Such instruments are no longer a rarity in the laboratory and are likely to become more common in the future as they offer new prospects in microscopy for biologists of widely varied interests, particularly those wishing to study the inter-relationships of plants and their associated micro-organisms.

The epifluorescence microscope, by virtue of its ability to transmit a choice of u.v. radiations through the objective lens directly on to the object, is ideal for examination of surfaces. Tissues of any thickness can be examined provided, of course, they can be suitably placed on the stage. No clearing is necessary for the purpose of assisting the passage of light, although the removal of chlorophyll with appropriate solvents may assist differentiation at certain wavelengths. Making the object visible depends on the presence of natural fluorochromic substances such as chlorophyll or lignified structures or, alternatively, fluorescent stains applied to the tissues or the micro-organisms.

(i) *Fluorescent stains and optical brighteners*

The selection of either the exciting wavelength, dichroic mirror system or the fluorescent stain depends on a variety of factors but the processes involved are generally simple and rapid. The optical brightening agents are generally useful (Paton & Jones 1971) but a variety of other useful fluorescent stains is available and eminently suitable for this application, and for most purposes a formulation containing acridine orange with an optical brightener gives excellent results. The optical brightener should be compatible with acridine orange in solution. A wide range of commercial brighteners is available but the most useful are those which are excited at relatively short u.v. wavelengths such as 330–380 nm. The presence of the optical brightener has an enhancing effect on the fluorescent emission of acridine orange, particularly at such short u.v. wavelengths. Formalin is included in the staining solution not only as a fixative but also to kill the cells of the micro-organisms, thus ensuring a satisfactory uptake of the optical brightener. A surface-active agent allows an intimate contact of the stain with the micro-organisms that is not otherwise possible on some plant surfaces.

A general purpose stain of the following formulation provides satisfactory results for epifluorescence microscopy.

Stock solution A

Acridine orange 0·1% (w/v), formalin 0·5% (v/v), Triton X (BDH, UK) 0·01% (v/v) in distilled water.

Stock solution B

Cationic brightener (Uvitex AN, Ciba Geigy, UK, or author) 0·5% (w/v) in ethanol.

The active component of 'Uvitex AN' is the *p*-toluene sulphonate salt of 1,1-bis (3*N*,5-dimethylbenzoxazole-2yl)-methine and is the most effective brightener so far recognized for microbiological purposes.

For use, 0·1 ml of stock solution A and 1·0 ml of stock solution B are added to 100 ml of distilled water. Immersion of plant material in this dilute stain will provide the required effect within 15 to 30 s. Prolonged staining should be avoided. The tissue is then washed entirely free of excess stain and mounted in water for examination with the fluorescence microscope using u.v. emission of appropriate wavelength. If chlorophyll or other autofluorescent substances are present and conflict with the fluorescence of the micro-organism, the use of a shorter wavelength will be of considerable benefit.

Suitable alternative optical brighteners are available but should be carefully selected from the very wide range prepared primarily for the paper and textile industries. Cationic brighteners are the most effective in the present context as they do not obscure the internal contents of the microbial cells. All cationic brighteners, however, are not adequately absorbed by micro-organisms. As a suitable and simple screening test, a rootlet fragment should be dipped in a weak aqueous solution of the brightener for 10–15 s, washed and examined by short wave epi-illumination. The uptake of the fluorescent stain by the plant components should be minimal and both bacteria and fungi should be intensely bright. Such an alternative reagent suitable for plant surfaces usable with or without acridine orange is 'Photine ACR' (Hickson & Welch Ltd, Castleford, Yorkshire, UK).

(ii) *Determination of viable populations*

For the determination of the viable microflora on a surface the use of fluorescein diacetate in a phosphate buffer at pH 7·5 to 8·0 as a mountant for unfixed material will be found to provide good results for most of the micro-organisms in that situation (Paton & Jones 1975). Observation of such specimens must be made within 30 min of preparation as the fluorescein released by the esterases will not be retained in the organisms but will gradually diffuse out into the mountant and the resulting general background fluorescence will obscure the desired detail. The esterases of the plant tissue will also be capable of acting on the fluorescein diacetate but except where contact is made with cut or damaged tissue, such a fluorescence is slow to appear. The reaction will not occur in the presence of fixatives or other chemicals capable of destroying the enzymes.

Only a portion, often very small, of the microbial cells observed on the plant are viable and an indication of the proportion may be obtained by treating the tissue with a short-wave brightener prior to mounting with fluorescein diacetate solution. Thus the total population, coloured pale blue, is indicated by using short wave (*ca.* 330–380 nm) u.v. and only the viable population, coloured yellow-green, appears when longer wavelengths (*ca.* 450–490 nm) are used. By changing the filter system either the total or the viable populations may be recognized in the same field. It must be remembered, however, that not all micro-organisms possess the necessary esterase for the release of the fluorescein and dormant cells may be slow to react.

(iii) *Preparation of specimens*

A few basic methods for the preparation of specimens can be employed. Fresh tissues, e.g. leaf or root fragments, are immersed in the acridine orange-brightener stain for up to 30 s, gently washed and examined with or without mounting in water under a coverglass. If the surface is irregular the thicker films of water under the coverglass can cause poor definition at the higher magnifications. Damage to pieces of tissue, e.g. by forceps or during staining and mounting, can be avoided by pressing the tissue on a short strip of plastic-backed adhesive insulating tape which can then carry the tissue through the processing to final observation under the microscope. Similar tape can be used effectively for the examination of fine root systems. The roots are floated on to the adhesive surface of the tape which is then carefully withdrawn from the water and gently warmed to dryness, when the tissues will adhere to the adhesive and be available for staining without disturbance.

Again, plastic adhesive tape can be readily used for impression films. In this case it is necessary to ensure that the tape or its adhesive is not autofluorescent. The tape is stained and examined after mounting under a coverglass with water. The results are usually a considerable improvement on those obtained using clear adhesive tapes (e.g. Sellotape) or cellulose acetate. The plastic backed adhesive tapes are vulnerable to solvents such as acetone and chloroform that are used as components of some fixatives. A safe alternative is 70% ethanol.

(iv) *Fluorescent protein tracing*

Fluorescent protein tracing has not so far been applied to the plant situation to any marked degree (Paton 1964; Nairn 1976). This lack of incentive appears to relate to the difficulties often encountered in the preparation of antisera of adequate specificity and staining quality. Any efforts made towards the production of suitable antisera will be well rewarded as this technique has a unique value for the identification of organisms in mixed populations, which is a common feature of the epiphytic flora. The procedures adopted for plant surfaces are identical to those generally recommended for immunofluorescent staining. Fixatives containing formalin or alcohol may be used but bleached specimens are to be avoided. Care must be taken to remove all traces of free fixatives or clearing agents before applying sera. When organisms are likely to be protected by slime or other exudates the normal staining and washing times should be extended. As with the fluorescein diacetate method the general microbial population can be revealed by superimposing a short-wave optical brightener after the immunological procedure.

The methods so far described for fluorescence microscopy involve fluorochromes unsuitable for ordinary transmitted light microscopy. Duplicate preparations may therefore be required for some purposes—one stained by conventional procedures, e.g. with aniline blue and phenol, and the other with a fluorescent stain. A future development is the likely use of dyes such as Evans Blue which is visible by ordinary light, but which also acts as a fluorochrome when using a specially selected u.v. emission around 546 nm in a dichroic system.

A relatively new fluorescing stain, 4,6-diamidino-2-phenylindole or 'DAPI', has been shown to have an affinity for DNA and can be used effectively for demonstrating nuclear structures. This specificity could offer an advantage over the optical brighteners and acridine orange in terms of the absence of undesirable background staining. When excited with light at a wavelength of 365 nm the DNA–DAPI complex fluoresces a bright blue with a wavelength at or greater than 390 nm, while unbound DAPI and DAPI bound to non-DNA material may fluoresce a weak yellow (Porter & Feig 1980).

(v) *Examination of seeds*

The methods which have been described are generally applicable to the detailed microscopic examination of seeds. The microflora of the outer surface can be revealed quickly by immersing the whole seed in the acridine orange-brightener stain for a few seconds, washing and mounting on Plasticine on a slide. The depth of focus problem already mentioned is more acute when working with seed; nevertheless, excellent differentiation is usually obtained. The focussing difficulty is partly overcome with larger seeds such as beans by pre-soaking and staining fragments of the testa. The internal surface of the testa is commonly more important, being in direct contact with the viable structures of the plant. The brilliant auto-fluorescence of embryonic and endosperm tissue tends to mask any imposed fluorescence and such tissues must be fixed in fluids containing non-fluorescent dyes: the subsequent fluorescent staining will then contrast with a duller background.

Further applications of such techniques to examine microscopically relatively large and optically dense natural materials to reveal their surface microflora *in situ*, were suggested by Paton & Jones (1971, 1975).

3. References

BEVEGE, D. I. 1968 A rapid method for clearing tissues and staining intact roots for detection of mycorrhizas caused by *Endogone* spp. and some records of infection in Australian plants. *Transactions of the British Mycological Society* **51**, 808–810.

BUTLER, E. E. & MANN, M. P. 1959 Use of cellophane tape for mounting and photographing phytopathogenic fungi. *Phytopathology* **49**, 231–232.

DICKINSON, C. H. 1967 Fungal colonisation of *Pisum* leaves. *Canadian Journal of Botany* **45**, 915–927.

DICKINSON, C. H. 1974 Impression method for examining epiphytic micro-organisms and its application to phylloplane studies. *Transactions of the British Mycological Society* **63**, 616–619.

DICKINSON, C. H. & PREECE, T. F. 1976 *Microbiology of Aerial Plant Surfaces*. London & New York: Academic Press.

FOWLER, A. M. & OWEN, H. 1971 Studies on leaf blotch of barley (*Rhynchosporium secalis*). *Transactions of the British Mycological Society* **56**, 137–152.

JANES, B. S. 1962 Leaf-clearing techniques to assist fungal spore germination counts. *Nature, London* **193**, 1099–1100.

NAIRN, R. C. 1976 *Fluorescent Protein Tracing* 4th edn. Edinburgh: Churchill Livingstone.

O'BRIEN, T. P. & VON TEICHMAN, I. 1974 Autoclaving as an aid in the clearing of plant speciments. *Stain Technology* **49**, 175–176.

PATON, A. M. 1964 The adaptation of the immunofluorescence technique for use in bacteriological examinations of plant tissue. *Journal of Applied Bacteriology* **27**, 237–243.

PATON, A. M. & JONES, S. M. 1971 Techniques involving optical brightening agents. In *Methods in Microbiology* Vol. 5A, ed. Norris, J. R. & Ribbons, D. W. pp.135–144. London & New York: Academic Press.

PATON, A. M. & JONES, S. M. 1975 The observation of micro-organisms in fluids using membrane filtration and incident fluorescence microscopy. *Journal of Applied Bacteriology* **38**, 199–200.

PORTER, K. G. & FEIG, Y. S. 1980 The use of DAPI for identifying and counting aquatic microflora. *Journal of Limnology and Oceanography* **25**, 943–948.

PREECE, T. F. 1959 A staining method for the study of apple scab infections. *Plant Pathology* **8**, 127–129.

ROVIRA, A. D. 1956 A study of the development of root surface microflora during the initial stages of plant growth. *Journal of Applied Bacteriology* **19**, 72–79.

ROVIRA, A. D., NEWMAN, E. I., BOWEN, H. J. & CAMPBELL, R. 1974 Quantitative assessment of the rhizoplane microflora by direct microscopy. *Soil Biology and Biochemistry* **6**, 211–216.

RUINEN, J. 1961 The phyllosphere; an ecologically neglected milieu. *Plant and Soil* **15**, 81–109.

Abstracts of papers read at meetings of the Society are published without verification of their scientific content.

Selected Abstracts of Papers Presented at the Summer Conference

Sequence Homology in *Rhizobium meliloti* Plasmids

P. DE LAJUDIE, L. JOUANIN, S. BAZETOUX AND T. HUGUET
Biologie Moléculaire Végétale, Associé au C.N.R.S.
No. 40, Univer, Paris-Sud, 91405—Orsay, France

An alkaline procedure, followed by a CSCl-EtBr centrifugation, was used for the isolation of plasmids from six symbiotically efficient strains of *Rhizobium meliloti* originating from different geographical locations and known to carry only one large plasmid (ranging from 90 to 145×10^6 Daltons). Each purified plasmid has been digested with eight restriction endonucleases: number of bands, molecular weight and band multiplicity were established. Plasmid DNAs yielded very complex cleavage patterns; only Kpn 1 and Xba I gave a limited number of bands. Examination of fingerprints revealed very few, if any, fragments of the same molecular weight common to the plasmids of these six strains, whatever the restriction enzyme used and the geographical origin of the strains. The absence of similarity between restriction patterns suggests that plasmid structure has not been highly conserved during evolution.

DNA-DNA hybridizations, however, revealed a significant extent of sequence homology between plasmids. Homologous sequences are likely to be scattered throughout the plasmid genome. Southern hybridization has demonstrated that one or a few restriction bands carry sequences homologous to all the other strains studied whatever their geographical origin. This sequence homology thus gives evidence for highly conserved sequence in *Rhizobium meliloti* plasmids. The biological function of this short conserved sequence (less than 4×10^6 Daltons) is unknown.

Induced Non-pathogenic Infection of Plant Cells

SRIYANI K. D. ALOYSIUS AND A. M. PATON
Division of Agricultural Bacteriology,
University of Aberdeen, Aberdeen AB9 1UD, UK

Following previous observations of the possible role of L-forms in bacterial plant disease, attempts have been made to induce selected organisms to enter and live symbiotically within plant cells. The successful first stages of this procedure are described where the L-forms of *Pseudomonas syringae* and *Beijerinckia indica* have been produced and then associated with single cell plant tissue culture suspensions. Occupation of the plant cells by these organisms was recognized by phase microscopy. Both types of cells appeared to accept the association and retained their viability. The implications of this procedure were briefly considered.

Site of Action of the Wildfire Toxin Produced by *Pseudomonas tabaci*

J. G. TURNER
School of Biological Sciences, University of East Anglia, Norwich NR4 7TJ, UK

Wildfire toxin was extracted and purified from culture filtrates of *Pseudomonas tabaci.* In the presence of ATP and Mg^{++}, the toxin inactivates glutamine synthetase (GS) from *Nicotiana tabacum,* the host of *Ps. tabaci,* as well as GS from eukaryote and prokaryote sources.

Tobacco-leaf GS was inactivated *in vivo* to <5% of basal levels 4 h after infiltrating the leaves with a solution of wildfire toxin. GS activity did not recover in these leaves which became chlorotic and necrotic. Ammonia increased over a 48-h period in tissues where toxin had caused near-complete GS inactivation. Necrotic symptoms occurred when intracellular ammonium ion concentrations reached 20 to 30 mmol/l.

Small changes in tobacco-leaf soluble protein, amino acids, chlorophyll and NAD-dependent glutamate dehydrogenase activity were detected after GS-inhibition of toxin-treated leaves. It is proposed that GS is the site of action of wildfire toxin, and symptoms are due to the subsequent accumulation of ammonia, the substrate of the inhibited enzyme.

Detection of Small Numbers of Phytopathogenic Bacteria Using the Host as an Enrichment Medium

Y. HENIS, Y. OKON, EDNA SHARON AND Y. BASHAN
Department of Plant Pathology and Microbiology, The Hebrew University of Jerusalem, Rehovot 76-100, PO Box 12, Israel

Detached leaves of tomato cv. VF-198 and pepper cv. Maor were surface-sterilized in 0·5% NaClO for 3 min, washed with sterile water, placed on 0·5% water agar and inoculated with 1·0 ml suspension of 10–1000 cells/ml of *Pseudomonas tomato* (tomato) or *Xanthomonas vesicatoria* (pepper). After incubation under fluorescent light at 24°C for 48–120 h, the leaves were again surface-sterilized and washed in sterile water. This procedure removed most of both bacterial pathogens and contaminants that developed on the leaf surface. The leaves were then homogenized in sterile water and aliquots of dilutions of the homogenate were plated on selective medium. After 48 h incubation typical fluorescent oxidase negative *Ps. tomato* or yellow *X. vesicatoria* colonies were counted. Bacterial counts increased significantly (10^5–10^7/g of leaves) inside the pathogen-inoculated leaves but not inside leaves inoculated with saprophytic *Ps. fluorescens*. Symptoms of bacterial speck of tomato and bacterial scab of pepper appeared in the detached, inoculated leaves after 5 d incubation in the Petri dishes. This method was successfully used to detect pathogens present in very small numbers in suspected commercial seed lots of tomato and pepper and in leaves from suspected fields.

Studies on the Bacteriophagy of Fluorescent Pseudomonads Isolated from Stone Fruit Trees in Portugal

J. M. S. MARTINS AND M. MARGARIDE F. MESQUITA
Estação Agronómica Nacional,
P-2780 Oeiras, Portugal

A collection of about 150 strains of fluorescent pseudomonads was isolated by the leaf-washing method from the leaves of one variety of apricot and six varieties of cherry trees in an experimental plot located at Alcobaça (Portugal). According to the 'LOPAT' scheme, 79% of these bacterial strains belong to Group Ia and 6% to Group Ib, the remainder being able to pectolyse potato, some of them producing levan-type colonies. Six strains were used to propagate 54 bacteriophages isolated from leaves and soil collected at the same plot. There is no evidence of specificity of any phage or group of phages to bacterial strains isolated from the same variety or species of host plants or sharing a similar spectrum of biochemical reactions. Nevertheless, the population of fluorescent pseudomonads epiphytic on cherry and apricot trees can be differentiated by the relative frequency distributions of their strains by classes of sensitivity to phages. Thus, fifty per cent of the 71 apricot strains are lysed by 10 to 30 phages, whereas 53% of the 87 cherry strains are either resistant to all or lysed by less than 10 phages, and 30% are lysed by more than 30. These data show that in this group of bacteria there is a certain kind of host specificity which is not evidenced by biochemical testing.

A Density Gradient Procedure for Separating Microbial Cells from Soil

N. J. MARTIN
Department of Microbiology, West of Scotland Agricultural College, Auchincruive, Ayr KA6 5HW
R. M. MACDONALD
Department of Soil Microbiology, Rothamsted Experimental Station, Harpenden, Herts. AL5 2JQ, UK

Non-filamentous micro-organisms were extracted from soil by density gradient centrifugation in silica sol + polyvinyl pyrrolidone (Percoll*) after centrifugal washing procedures. Since Percoll has an osmolality <20 mOs/kg H_2O, its osmotic pressure may be readily adjusted and micro-organisms float at their true buoyant density in the gradient. Percoll is non-toxic and has a viscosity of 10±5 cP, hence micro-organisms may be rapidly banded isopycnically in a physiologically active condition.

Preliminary experiments showed that most soil micro-organisms had a density range 1·081–1·123 g/cm^3 while *Rhizobium* isolated from crushed root nodules on Percoll was split into two bands of densities 1·081–1·110 and 1·041–1·073 g/cm^3. The lighter cells were the more pleomorphic.

In the case of soil suspensions, only micro-organisms which were desorbed from soil particles were extracted on to density gradients of Percoll. These cells normally constituted *ca.* 10% of those detected in soil homogenates by microscopy. However, 71% of *Serratia marcescens* was recovered by plating Percoll when the cells were initially added to the soil homogenate. Electron microscopy of soil micro-organisms isolated by this method showed an unusual range of surface ornamentations on cells of bacterial dimensions.

Yield Increases on a Commercial Scale in Grasses Inoculated with *Azospirillum* In Israel

Y. HENIS, Y. KAPULNIK, S. SARIG, I. NUR AND Y. OKON
Department of Plant Pathology and Microbiology, The Hebrew University of Jerusalem, Rehovot 76-100, PO Box 12, Israel

Various grasses were inoculated with nitrogen-fixing bacteria of the genus *Azospirillum* under field conditions in 100–200 m^2 plots in the northern Negev and Beth Shean Valley in Israel. In both irrigated and non-irrigated soils, inoculation significantly increased the yield of components of commercial

* Percoll is a trademark of Pharmacia Fine Chemicals, Uppsala, Sweden.

value as follows (%): *Zea mays* cv. Jubilee sweet corn ear yield, 16; cv. H-Nansi, forage dry matter, 24 and total N-yield, 72; cv. H-851, corn meal dry matter, 11; *Sorghum bicolor* cv. H-226, top yield, 35; *Panicum miliaceum*, seed yield, 13; *Setaria italica*, forage crop (total plant weight), 45 and total N-yield, 128; *Triticum turgidum* var. *durum* cv. Inbar, grain yield, 10; *Triticum aestivum* cv. Barkai, total plant weight, 17; cv. Miriam, grain yield, 20. It was concluded that inoculation of summer and winter cereal crops in Israel may save valuable nitrogen fertilizer.

Herbicide Effects and Fate in the Rhizosphere of Wheat

P. J. MUDD*, R. J. HANCE**,
M. P. GREAVES** AND S. J. L. WRIGHT*
**School of Biological Sciences, University of Bath, Bath BA2 7AY*
*and **ARC Weed Research Organisation, Yarnton, Oxford OX5 1PF, UK*

The effects of the urea herbicide isoproturon on microbial populations and processes in the rhizosphere of wheat and in plant-free soil were examined under field, glasshouse and laboratory conditions. Degradation of isoproturon (^{14}C-labelled) was also studied.

In field experiments, isoproturon applied to soil at the recommended rate (2·5 kg active ingredient/ha) caused variable effects including temporary increases and decreases in bacterial and fungal propagules, but no changes in levels of NH_4^+-N, NO_2^--N, NO_3^--N or PO_4^{3-}. Similar results, without distinct trends, were recorded using 0·75 kg isoproturon/ha with glasshouse-initiated plants subsequently transferred outside although in this case the response of fluorescent pseudomonads differed in the rhizosphere and root-free soil. Growth in liquid media of pure cultures of fungi and bacteria isolated from the wheat rhizosphere was apparently unaffected by isoproturon at levels equivalent to normal and 10× field application rate. However, at levels equivalent to normal, 10×normal and 10×field rate in agar media changes occurred in the growth of several fungi, including the pathogen *Gaeumannomyces graminis*. The changes included increases and decreases in radical growth, mycelial density, sporulation and pigmentation. Under glasshouse conditions, isoproturon was degraded more rapidly in soil planted with wheat than in unplanted soil. Degradation was relatively faster the lower the initial herbicide levels and was favoured by elevated temperature and moisture content. Isoproturon was not significantly bound to the human fraction and its degradation pathway was the same in planted and unplanted soils.

Organic Secretions by Duckweed and their Relationship with Epiphytic Bacteria

J. H. BAKER AND I. S. FARR
FBA River Laboratory, East Stoke, Wareham, Dorset BH20 6BB, UK

Duckweed (*Lemna minor* l.) supports approximately 10^7 bacteria/cm on the lower surface of mature plants (Hossell, J. C. & Baker, J. H. 1979 *Journal of Applied Bacteriology* **46**, 87–92). Some aquatic macrophytes are known to secrete dissolved organic carbon (DOC) compounds into the surrounding water. The purpose of this investigation was to discover whether *L. minor* was such a plant and if so how much DOC is produced and whether it could be utilized by the epiphytic bacteria. The bacteria might affect DOC production in two ways; first if they utilize the DOC for their own growth the apparent DOC production rate in their presence would be less than in their absence; second they might stimulate the plant to produce more DOC possibly through hormonal production. Hence it was necessary to determine DOC production rates in the presence and absence of epiphytic bacteria.

Axenic *L. minor* was obtained using a combination of antibiotics and u.v. light. It was cultured in the laboratory in autoclaved natural water and growth was estimated as the increase in dry weight per unit time. DOC was measured by an automated u.v. photolysis method (Baker, C. D., *et al.* 1974 *Freshwater Biology* **4**, 467–481). Approximately 2% of the carbon fixed by *L. minor* appears to be secreted as DOC. Analysis of the DOC by ultrafiltration shows that much of it is of relatively low molecular weight. The measured DOC production rate in the presence of epiphytic bacteria was generally lower than that determined for axenic *Lemna*.

Digestion of Plant Particles by Rumen Phycomycete Fungi

C. G. ORPIN AND YVONNE HART
Agricultural Research Council, Institute of Animal Physiology, Babraham, Cambridge CB2 4AT, UK

The anaerobic rumen phycomycete fungi *Neocallimastix frontalis, Piromonas communis* and *Sphaeromonas communis* were grown in pure culture *in vitro* using plant particles as sole carbon and energy sources. The plant particles used were perennial rye-grass, hay (mixed sward) and wheat straw leaves. The dried plant tissues were milled and seived, and the fraction passing a 2 mm but not a 1 mm mesh were retained and incorporated at a rate of 10 mg dry wt/ml into an anaerobic liquid culture medium deficient in metabolizable substrates. Media were inoculated with 2 ml of zoosporecontaining supernatant fluid from cultures grown on the same substrate, and incubated at 39°C for 4 d. The residual particles were analysed by a modification of the methods of Thornber & Northcote (Thornber, J. P. & Northcote, D. H. 1962 *Biochemical Journal* **82**, 340–346) and their composition compared with uninoculated particles.

The extent of digestion of water-insoluble components varied with the fungal species and the plant species, but the most extensive digestion (45% loss of dry wt) was shown by *P. communis* with perennial rye-grass. Under the same conditions, *N. frontalis* and *S. communis* digested 42 and 31% of the dry wt respectively. Corresponding figures for hay were 34, 32 and 29% and for wheat straw leaves, 37, 35 and 30%, respectively. Partial degradation of the pectin, xylan, hemicellulose I and II and cellulose fractions occurred. The extent of digestion varied with the fraction concerned, but quantitatively the most significant digestion was of hemicellulose I and cellulose. For example, wheat straw leaves contained 29·6 and 24·5% dry wt as hemicellulose I and cellulose respectively. The digestion of these fractions by *N. frontalis, P. communis* and *S. communis* was, respectively, 52·4 and 58·1%; 51·8 and 50·4%; and 50·8 and 39·7%.

Growth of all three species on plant particles resulted in a loss of part of the lignin fraction from the particles. With wheat straw, *N. frontalis* digested or released from the tissues 19·4% of the lignin fraction, *P. communis* 21·9% and *S. communis* 16·4%. The fate of the lignin fraction has yet to be determined.

Growth of all three isolates on the particles resulted in a diminution in the size of the particles, which was greater with *P. communis* and *N. frontalis* than with *S. communis.*

Survival and Growth of *Bacillus cereus* in Cooked Rice

K. NEWTON, A. SEAMAN AND M. WOODBINE
Microbiology Unit, Faculty of Agricultural Science,
University of Nottingham, Sutton Bonnington LE12 5RD, UK

Bacillus cereus NCTC 2599 inoculated into boiled rice grew at temperatures between 22 and 44°C, but not at 10 or 50°C. The addition of beef or chicken protein gave higher numbers, especially at 22°C. The addition of 5% NaCl to the water in which the rice was boiled depressed growth at 22°C but had no effect at higher temperatures. Rice fried in corn oil and supplemented with 10% beaten egg gave higher levels at 22°C than with the other protein supplements, but at higher temperatures there was little difference between the effect of the protein supplements.

Mannitol-egg yolk-polymyxin-phenol red agar was used for the isolation of *B. cereus* from rice and rice dishes.

A commercial rice sample had a presumptive *B. cereus* count of $5{\cdot}2 \times 10^2$/g. After boiling for 20 min this was reduced to 56/g. Storage for 24 h of the boiled sample gave 63/g whereas storage at room temperature gave 9×10^4/g. This was reduced by frying to $1{\cdot}7 \times 10^3$/g.

Samples from four commercial food retailers showed *B. cereus* levels ranging from 60/g to 2×10^4/g with the highest levels in mixed foodstuffs (chicken and vegetable rice risotto and cold boiled rice in a salad).

Production of Microbial Phytase

M. HUSSEIN, R. MAHMOUD, E. RAMADAN,
I. RIZK AND A. GIBRIEL
*Food Science Department, Ain-Shams University,
Shobra-Khaima, Cairo, Egypt*

Nutritional and environmental conditions for microbial phytase production by *Bacillus licheniformis, B. subtilis, Candida intermedia, Saccharomyces cerevisiae, Glomerella cingulata* and *Aspergillus* sp. were studied. The presence of wheat bran in the basal media as a source of phytic acid increased the microbial enzyme production. Ammonium sulphate proved to be the most preferred nitrogen source for phytase production by yeast and *B. subtilis,* while ammonium chloride and sodium nitrate were the best in the cases of *B. licheniformis* and fungal strains, respectively. The optimum initial pHs for the enzyme production were 3, 4 and 5 for yeast, fungal and bacterial strains, respectively. The optimum incubation temperatures were 28, 32 and 37°C for fungal, yeast and bacterial strains, respectively. Three days incubation period is highly recommended for the phytase production from bacterial and yeast strains and 5 d for fungal strains.

The Taxonomic Position of *Erwinia rhapontici*

M. B. SLADE AND A. I. TIFFIN*
Department of Immunology, Royal Postgraduate Medical School, Du Cane Road, Hammersmith, London W12 and **Department of Microbiology, University of Reading, London Road, Reading RG1 5AQ, UK*

Fifty-eight strains of *Erwinia* were grouped by numerical taxonomy according to 107 different tests using the simple matching coefficient, S_{sm} (Sokal & Mitchener 1958 *University of Kansas Science Bulletin* **38**, 1409) and group average cluster analysis.

Six strains of *Erwinia rhapontici* were found to link with the 'herbicola group', not the 'carotovora group' in which the species is currently placed in Bergey's Manual. Three of the *E. rhapontici* strains were isolated from the epiphytic flora of pear (*Pyrus communis*) or hawthorn (*Crataegus* sp.), to the species is similar to *E. herbicola* in that it is a member of the epiphytic flora. Three of the *Erwinia* strains formed a new group (E1) related to both *E. rhapontici* and *E. herbicola.* On the basis of these observations we suggest that *E. rhapontici* should be placed in the 'herbicola group' of *Erwinia.*

The cluster analysis divided *E. herbicola* strains into four clusters—two of which correspond to the *E. herbicola* var. *herbicola* and *E. herbicola* var. *ananas* clusters described by Goodfellow *et al.* (1976 *Journal of General Microbiology* **97**, 219). One of the two previously undescribed clusters within *E. herbicola* contained non-pigmented strains, demonstrating that they are taxonomically distinct from the pigmented *E. herbicola* strains. The fourth cluster, characterized by a low (36°C) maximum growth temperature, contained three pigmented strains which were originally named *Agrobacterium gypsophilae* by Maas Geesteranus.

The Taxonomy of *Pseudomonas* Species Associated with Diseases of the Cultivated Mushroom

B. A. UNSWORTH AND T. F. PREECE
Plant Sciences Department, Agricultural Sciences Building, University of Leeds, Leeds LS2 9JT, UK

Two hundred isolates of fluorescent pseudomonads from healthy and diseased mushroom caps have been examined in a search for characters that are useful in distinguishing the pathogenic species *Pseudomonas tolaasii* and *Ps. agarici* from apparently saprophytic species and in the taxonomy of mushroom pseudomonads in general. The main points that have emerged so far can be summarized as follows:
(i) *Pseudomonas tolaasii* and *Ps. agarici* are distinct from each other and from other fluorescent pseudomonads. (ii) There are several distinct groups of saprophytic fluorescent pseudomonads on mushroom caps. (iii) Carbon source utilization tests have proved useful in identifying the various groups. (iv) *Pseudomonas tolaasii* does not fall into biotype II of *Ps. fluorescens* as stated in Bergey's Manual, 8th edition; rather it appears to belong either in biotype G of Stanier *et al.* (1966) or perhaps as a distinct, new biotype. (v) *Pseudomonas agarici* is unlike any other fluorescent *Pseudomonas* species.

These points were discussed and the findings of a preliminary numerical taxonomic study of the mushroom pseudomonads were presented.

Subject Index